3

THE PLAN OF CAMPAIGN 1886-1891

SUPPLEMENT GIVEN AWAY WITH THE **WEEKLY FREEMAN** 30TH JULY, 1887.

SOLD AGAIN.

Our old "Friend," the Irish Landlord, believing he has frightened the wits out of Farmer Pat, proceeds to pick his pockets, and finds that all that [illegible] person carries about him is the PLAN OF CAMPAIGN!

THE PLAN OF CAMPAIGN

1886-1891

Laurence M. Geary

CORK UNIVERSITY PRESS

First Published 1986
Cork University Press, Cork, Ireland.

British Library Cataloguing in Publication Data

Geary, Laurence
The plan of campaign, 1886-1891
1. Land tenure—Ireland—History—19th century
2. Land tenure—Political aspects—Ireland

I. Title
33.3'23'0945 HD625

ISBN 0 902561 43 X

Printed in Ireland by Reprint Limited
and bound by Duffy Bookbinders Limited.

Introduction

The Irish land question, unlike the poor, has not always been with us. It merely seems that way. Equally, its appeal for scholars, on both sides of the Atlantic, is perennial and seemingly inexhaustible. In recent years, it has been subjected to fresh scrutiny and a new generation of revisionist historians has shattered many of the old shibboleths and generalisations concerning post-Famine changes in agriculture and landlord-tenant relations generally. But, as a leading authority on the subject has observed, 'historians almost always have problems with causation'.[1] Consequently, if new orthodoxies tended to develop, and particularly if they acquired a certain modishness, their attempted demolition became a virtual *sine qua non*.

The post-Famine land system, as conventionally portrayed, was intolerable and rapacious, with rack-renting, wholesale evictions and endemic agrarian crime as the norms. A severe agricultural depression in the late 1870s exacerbated the situation and culminated in the land war of 1879.[2] The traditional interpretation of the land war as a spontaneous mass uprising of a homogeneous, rural population against predatory landlordism has been radically reinterpreted in recent years.

One novel suggestion was that the agitation was sparked off by a 'frustrated revolution of rising expectations'. According to James Donnelly, the quality of post-Famine rural life improved immeasurably in the wake of unprecedented agricultural prosperity. However, the agricultural depression of the late 1870s stymied the expectation that living standards would continue to improve and, as a result, large numbers of tenant farmers rose in revolt in order to protect their interests.[3]

At best, the 'rising expectations' thesis won only partial acceptance with Marxist historians. Paul Bew accepts the validity of Donnelly's argument when applied to the more prosperous tenant farmers of the midlands, east and south-east, but contends that it is inapplicable to the subsistence tenantry of the west who took the lead in the Land League. From this viewpoint, the land war was a class struggle fought at two levels—between landlords and tenants and by tenants among themselves. Tensions and conflicts of interest had developed between the poorer and the more prosperous farming classes, to say nothing of those that existed between every kind of tenant and the landless labourers. As the Land League expanded these strains became more pronounced and determined the strategies adopted during the course of the land war.[4]

While Bew emphasises rural class tensions, Samuel Clark argues in favour of communal solidarity. In his view, the economic depression of the late 1870s brought disparate urban and rural social groups together in a mass movement which aimed at the abolition of landlordism. A common set of interests forged an enduring collectivity, many of whose members mobilised in its defence.[5]

David Fitzpatrick believes that the primary agent of rural unrest was not class but family structure. He contends that conflicts apparently between classes may often be interpreted as struggles within families; conflicts within classes could be argued to be struggles between family factions.[6]

The joint editors of a recent collection of essays on rural collective action acknowledge the role of family structure in agrarian agitation but, unlike Fitzpatrick, feel it unnecessary to promote it at the expense of agrarian classes. In their opinion, several agencies contributed to agrarian unrest, including family, class, religion, territorial divisions and peer groups. 'Few active collectivities', they conclude, 'are based entirely on one type of social structure; most are a complex result of the divergent or reinforcing effects of several different sources of cohesion and cleavage'.[7] W. E. Vaughan, rather more engagingly, posted a similar caveat: while 'the neatly packaged monocausal explanation is intellectually attractive... the events studied usually demand an untidy eclecticism'.[8]

The Plan of Campaign, which was promulgated in the Parnellite weekly, *United Ireland,* on 23 October 1886, was the second phase of the Irish land war. Like the Land League before it, the Plan was spawned by economic difficulties. During the winter of 1885-6, tenants began to withhold rents and to combine on individual estates in order to pressurise landlords into granting abatements. Such tactics were emphatically condemned by Parnell, on the grounds that they retarded political progress towards Home Rule. The political upheavals during the summer of 1886 and, subsequently, the disinclination of the newly elected Conservative government to tackle the deteriorating economic crisis in Ireland made a renewed agrarian agitation during the winter of 1886 inevitable.

The Plan of Campaign was not the official policy of either the National League or the parliamentary party. Parnell was at pains to distance himself from the agitation and publicly disavowed it on several occasions. However, its leadership and principal exponents were drawn from the agrarian wing of the party and the secretary of the National League, who drafted the Plan's manifesto, explicitly pledged the League's support. At local level, the two organisations generally shared a common leadership.

The Plan was more widely adopted than has hitherto been accepted. John Dillon, whose figures were never subsequently questioned by historians, underestimated the scale of the agitation by seventy-five per cent. The vast majority of cases occurred in the poorer western half of the country. This is not to suggest that the agitation was simply the reaction of an impoverished peasantry to adverse economic circumstances. The Plan of Campaign was much more complex than that and its many intricacies accurately reflected the grievances and aspirations of a far from homogeneous community. It is impossible to calculate the total number of tenants involved, ranging as they did from 10 on the Delmege property in County Limerick to 5,000 on Lord Dillon's 80,000-acre Mayo estate. But, without exception, they were Catholics and the Plan of Campaign was an exclusively Catholic movement. Ireland's two best-known churchmen gave it their archiepiscopal blessing; the other bishops, with few exceptions, were favourable; the parochial clergy figured prominently in the local leadership. This ubiquitous clerical involvement and the absolute identification of Irish nationalism with Catholicism in the wake of Gladstone's abortive attempt to introduce Home Rule in the summer of 1886 alienated Protestant tenants and minimised the impact of the Plan of Campaign on north-east Ulster. For a time, economic interests had superseded religious affiliations in the north and Ulster Protestant tenants had joined the Land League in significant numbers. Such interdenominational political co-operation was, however, short-lived; it disintegrated finally in the political upheavals of 1885-6, when agrarian radicalism was sacrificed to sectarian conflict.[9]

The Plan of Campaign was seen by its opponents as an anarchical attack on sanctity of contract and the rights of property and was particularly offensive to a Tory government publicly committed to twenty years of resolute rule in Ireland. The timidity of the landlords' individual and collective response infuriated Sir Michael Hicks Beach and Arthur Balfour, successive Irish Chief Secretaries. The widely publicised, co-incidental condemnation of the Plan of Campaign by the Pope and by Parnell in mid-1888 weakened the agitation less than its constant shortage of money, for which the relief of a large and ever-increasing number of evicted tenants was primarily responsible.

The penultimate phase of the struggle centred on the reverberations surrounding the landlord syndicate formed at the instigation of Arthur Balfour to defeat the Plan of Campaign on the east Cork estate of C.W.T. Ponsonby, a retired English naval officer. The ensuing conflict drained the Plan's resources and, in the winter of 1890, a six-man delegation embarked for the United States on a fund-raising mission. They had scarcely touched on American soil before their hopes, and those of 1,600 evicted tenants in Ireland, were dashed by the outcome of the O'Shea divorce case. The subsequent rupture in the Parnellite ranks marks the end of the Plan of Campaign as a vital, hortatory movement. The majority of the extant struggles collapsed in 1891, although some dragged on until early in the twentieth century and one notorious case was not finally settled until 1913.

After the rejection of Gladstone's Home Rule Bill in the summer of 1886, the Irish once again secretly changed the question and the Plan of Campaign became, in effect, the Irish Question. The agitation dominated the closing years of the decade, provoked as much debate and raised as many contemporary hackles as had the Land League before it but, whereas the first phase of the Irish land war has been microscopically probed and analysed, the Plan of Campaign has been treated cursorily by historians. Aspects of the agitation have attracted scholarly attention, such as the role of John Dillon,[10] the response of the Salisbury government,[11] the attitude of the papacy and of the Irish Catholic hierarchy.[12] Individual conflicts have been researched,[13] while James Donnelly devoted a chapter of his pioneering work on County Cork to the Plan of Campaign.[14] Irrespective of their individual worth, these studies are necessarily incomplete and portray one-dimensionally an agitation of some complexity. There has long been the need for a fully-rounded history of the Plan of Campaign and it is hoped that the present work will meet that need.

The sources used are mainly archival, especially estate records and landlords' private papers, the mortgage ledgers of the Representative Church Body and of Maynooth College, the records of the Irish Land Commission and the papers of Irish and British politicians. The State Paper Office in Dublin Castle is a virtual cornucopia for the researcher, housing, as it does, the minutiae of administrative and constabulary response to Irish agitation. Parliamentary commissions of inquiry into the land system yielded insights into individual estates. Newspapers occasionally contained incisive editorial comment but, more significantly, they provided the day-to-day information necessary to the understanding of any agitation.

1 *Land and Politics*

> The great and all absorbing topic in Ireland...is the land question. The land question is at the root of all other questions. The land is the only dependence of the people.
>
> Charles Stewart Parnell, speech at Kildare, 10 October 1885, *Freeman's Journal,* 12 October 1885.

On 30 May 1885, the *Cork Examiner* announced that the Duke of Devonshire had granted a twenty per cent rent reduction to his Cork and Waterford tenants. The report noted the 'great depression in the prices of stock and butter' and concluded that Irish farmers urgently needed a 'helping hand'. In October and again in December, the cabinet was advised of the great fall in agricultural prices and of the gravity of the economic situation generally.[1] In January 1886, Parnell warned the members of the House of Commons that conditions in Ireland were very serious and that the tenants were desperate.[2]

The Irish economy, following a slow recovery from the catastrophic depression of the late 1870s, plummetted once again in the mid-1880s. By 1886, agricultural production had fallen to seventy-nine per cent of its 1881 level[3] and, compared with the average for the four preceding years, the value of livestock and crops fell by eighteen and one half per cent in 1885 and 1886. Poor weather, soil exhaustion and increased foreign competition were primarily responsible. The falling agricultural prices were accompanied by a general restriction of credit on the part of banks, shops and money lenders.[4]

In general, tenants in the poorer western half of the country were more seriously affected by the depression. The tenants of Lord Clanricarde at Woodford, County Galway, were close to despair:

> Our corn crop is this year much below the usual average and the prices far less than they have been for years past. Our potato crop which is the chief support of most of us is in many portions of this district almost a complete failure. The great reduction in the price of stock and the utter impossibility of being able to sell our mountain cattle at any price have brought many of us to the verge of ruin.[5]

The Donegal tenants of Captain A. Hill claimed it was utterly impossible to sell 'small stock.. . .even at a sacrifice'.[6] The English liberal, Wilfrid Scawen Blunt, who was to play a conspicuous role in the later land agitation, claimed in April 1886 that a large part of the west of Ireland had ceased to be a money economy. Blunt, who had spent the previous month touring the remoter areas of Connaught, claimed that the tenants were impoverished and heavily indebted to banks and shopkeepers. 'There is no money anywhere, except in a very few hands, and the medium of exchange has already ceased in the poorer villages to be money at all. In Roscommon I found peasants, three days ago, paying for meal and other necessities with eggs, reckoned at two to the penny.'[7]

For many tenants even the judicial rents[8] had become rack rents, a point conceded by the rent commissioners who, from the end of 1885, began to fix the judicial rents at a level of ten to fourteen per cent below that which had prevailed for the previous four years.[9] In March 1886, the Inspector General of the Royal Irish Constabulary, emphasising the severity of the agricultural depression, wondered how the tenants had been able to pay any rent.[10] In the same month, the influential agricultural economist, Sir James Caird, wrote in the *Times* that the ratio of small farmers to comparatively large ones in Ireland was 4½ : 1 and he suggested that from the lands held by the former 'any economical rent has for the present disappeared'.[11]

Faced with this alarming situation, the tenants invariably petitioned their landlords to reduce the customary rents. These appeals, which were usually orchestrated by the local clergy, were almost always respectful and invariably stressed that the great depression in trade and agriculture was their sole motivating factor.

In December 1885, Father Coen and some 300 of the Woodford tenants signed a memorial requesting a twenty-five per cent reduction from Lord Clanricarde. They emphasised that they were not attempting to renege on their contractual obligations and promised that those who were able would do all in their power to fully discharge their debts.[12] Captain Hill's tenants asked for a twenty-five per cent reduction on the judicial rents and fifty per cent on the non-judicial ones and begged him not to increase their indebtedness by taking legal proceedings against them.[13]

Many landlords responded to their tenants' difficulties with prompt and generous concessions. Inspector General Reed of the Royal Irish Constabulary noted that 'landlords have shown great forbearance and tenants in general have shown a willingness to pay upon receiving some abatement'.[14] Many others, however, were either unwilling or unable to reduce their rents and relations between them and their tenants deteriorated rapidly. Lord Carlingford, an expatriate Irish Protestant landlord, who visited Ireland in October 1885, felt as if he had landed in a foreign and enemy country. 'The very air', he wrote, 'seems charged with hostility and hatred towards England and towards Irishmen also of my class'.[15] The landlords were universally condemned for their obduracy. The *Nation* indicted them for being 'invincibly ignorant' and claimed that Bishop Berkeley's description of them as 'men of vulturnine beaks with bowels of iron' still applied.[16] Parnell accused them of trying 'to draw blood out of a stone' and warned that they were standing on the edge of a precipice and were doing their best to get themselves pushed over.[17] A government official, E. G. Jenkinson, was equally critical and, as early as September 1885, had advised the cabinet that 'it would take very little trouble to get up a strong agitation against them'.[18]

The more radically outspoken of Parnell's followers, Tim Healy, John Dillon and William O'Brien, seemed determined to provoke just such an agitation. According to Healy, the 'No Rent Manifesto' of 1881-2 had failed for two reasons: the essential unity of the movement was broken when the rich farmers, fearing eviction, paid their rents and thereby left the poorer tenants at the mercy of the landlords; secondly, the Land League did not have sufficient funds to cover all expenses connected with evictions. In Healy's view, if a tenants' combination were to succeed, it would need to be adequately financed and thus be capable of surviving desertion by the self-interested.[19]

On 30 August at Killarney, Healy advised the tenants to combine and bank their rents, less a reduction, in the names of two or three trustees. Fear of forfeiting their deposits would prevent disloyalty and backsliding by 'sleeveens' and renegades. Should the landlord refuse to reduce rents, the money deposited would be used to fight him and to help the evicted. The landlord, having no resources to fall back on, would be unable to resist the demands indefinitely. If the tenants acted otherwise, warned Healy, they would make a laughing stock of themselves, set a bad example for the other tenants and encourage landlords generally to refuse reductions.[20]

A few weeks later in County Monaghan, Healy unveiled an even more radical scheme. Instead of combining on individual estates and forming 'mere parish organisations' which could easily be defeated, Healy advised the tenants to ignore their landlords and pay their rents into a common fund. 'They would then have a sum of six or seven or ten millions of money as a campaign fund, a war chest to fight the battle with'. Such a scheme, he suggested, would quickly bring the landlords to their knees.[21]

At a meeting of the central branch of the National League on 20 October, Dillon advised the tenants to 'lodge whatever rents they are able to pay...in the hands of trustees to be used as a campaign fund against the landlords'.[22]

In a caustically worded editorial in *United Ireland,* William O'Brien was extremely critical of the passivity of the majority of Irish tenants. He warned that landlords, backed up by the law, would always be too strong for individual and unorganised tenants and would only yield to combinations:

> If parish after parish banked its rent at the abated rate in trustees' names and pledged itself to use this war chest to succour every man attacked...a fund of a million of money would very quickly ...give a death blow to the delusion that landlording is the only trade in the country which must not suffer from depressed times.

He concluded with an almost biblical warning to the landlords that it would be better for even the most impoverished of their number 'to have eaten dry bread and gone barefooted and misapparelled during this disastrous year than have brought about by a refusal of just concessions the wrath and resolution of an angry people.'[23]

As the winter, and the gale day, of 1885 approached, the temperature of agrarian resistance began to increase perceptibly. Nationwide, tenants heeded the advice of Healy, Dillon and O'Brien and pledged themselves to combinations. Strongly worded resolutions, demanding the expulsion of traitors and backsliders from the League, were commonplace. The Kenmare tenants demanded the 'ignominious' expulsion from the League of any tenant who paid his rent secretly, 'as we consider the man who pays the least rack rent is offering a grievous wrong to every other class of the population whose demands, according to every law human and divine, deserve to be considered before those of any landlord'.[24] The by now classic agrarian speeches condemning land grabbing and advocating boycotting were increasingly heard.[25]

Irish loyalists, in particular the landowning classes, felt besieged. Rent strikes, sporadic violence, general lawlessness, extensive boycotting and a sustained attack by the radical nationalist press gave them good cause for alarm during the autumn and winter of 1885-6. Lord Salisbury's caretaker government, in an attempt to woo Parnellite support, introduced a land purchase measure and allowed the coercive legislation of 1882 to expire early in July 1885. The general election at the end of the year failed to produce a majority for any single party. Parnell held the balance of power and the Tories remained in office with his support. The loyalists felt that their traditional allies had abandoned them to a state of anarchy in Ireland.[26]

Early in 1886, a rather hysterical outburst from Lord Lansdowne's Kerry agent, J. Townsend Trench, appeared in the *Times*: 'an extraordinary and criminal combination... confiscation... organized resistance to payment of rent... civil war... scenes of violence... a disgrace to any civilized country'. He stressed that the existing law was totally inadequate to deal with the maelstrom in Ireland and demanded the re-introduction of coercive legislation. 'We are at a grave crisis', he concluded, 'and the interests of civil order, of trade, industry and prosperity in Ireland need prompt evidence that English honour and imperial unity are not to be sacrificed on the altar of party politics.'[27]

A fortnight later, immediately prior to the opening of the new parliament, deputations from the Irish Defence Union and the Irish Loyal and Patriotic Union met the Prime Minister in London. The *Times* described them rather fancifully as 'men belonging to every creed and class among the Irish people except the Jacobins and their peasant dupes'.[28] The Earl of Pembroke, Earl Fitzwilliam, Viscount De Vesci, A. H. Smith-Barry and Walter Morrison, each of whom was to play a leading role in advocating the landlord cause during the Plan of Campaign years, were in attendance. The anarchic state of the country was again stressed. 'The law is powerless... life is not safe, property is destroyed, and rents and debts cannot be collected... a fearful system called boycotting prevails'. They demanded that the law be enforced and, if necessary, strengthened. The *Times* concurred, indicating that coercion was the only antidote to the 'whirlpool of revolution' in Ireland.[29]

These sentiments were grist to Salisbury's mill and strengthened his determination to take a firmer line on Ireland. It was announced in the Queen's speech that the government was thinking of introducing coercion and thenceforth its days were numbered. The Tories' brief flirtation with the Parnellites ended on 28 January when the government resigned, having been defeated on Jesse Collings' celebrated amendment to the address.

The period of the caretaker ministry was marked by considerable speculation and debate on the question of Irish Home Rule. With the return of the Liberals to power and Gladstone's Pauline conversion, Home Rule became the central issue in Anglo-Irish relations. The main plank of the Irish party was national self-government with everything else, including the land question, playing a secondary role.

This policy was re-iterated by Parnell in August 1885 when he appealed for prudence and moderation in Ireland during the coming winter.[30] Parnell's dilemma was how to distance himself and his party from the agrarian agitation, with its concomitant hazards, so as to portray his movement as constitutional and law abiding. On 21 January 1886, he drew the attention of the members of the House of Commons to the seriousness of affairs in Ireland. He admitted that tenants were combining to resist the payment of customary rents but claimed that these movements were spontaneous and had received neither encouragement nor financial assistance from the National League. He declared that both he and his party were doing all in their power to stamp out boycotting and curb anti-rent combinations.[31]

At an informal meeting of the parliamentary party in the House of Commons on the following day, it was agreed that members should attend as few demonstrations as possible in Ireland and that they should moderate their language at those meetings they felt compelled to attend. Also, they would, in future, decline to advise tenants on the question of rent abatements.[32]

Moderate nationalist opinion supported the aims and methods of Parnell and the parliamentary party. In February 1886, the Catholic hierarchy stressed that, while the land question required an urgent solution, Home Rule 'alone can satisfy the wants and wishes as well as the legitimate aspirations of the Irish people'.[33] The influential *Freeman's Journal* tacitly concurred. An editorial on the bishops' statement opined that their 'position, ripeness of judgement, unimpeachable integrity and knowledge of the people entitle them to speak with the fullest authority'.[34]

Parnell's control over the National League was absolute[35] and he was determined to curb the excesses of maverick branches which, by their actions and resolutions, were embarrassing the national movement. At the end of January 1886 he informed Gladstone, through Katherine O'Shea, that those branches which would not 'keep within bounds' would be dissolved.[36] Shortly thereafter, the organising committee of the League withdrew grants from evicted tenants in perpetually disturbed districts. 'They wish,' wrote Tim Harrington, the League's secretary, 'to save the general organisation from even the suspicion of sending funds to places where outrages of this kind have been occurring and they regard this course as necessary for the safety and character of the organisation at the present time.'[37] In June, Harrington declined to give an opinion on the question of boycotting as it would connect the League with this policy and they were determined to avoid 'any charge of this character that may be preferred against them by the enemies of the national movement'.[38] The unfortunate president of the Abbeydorney branch, near Tralee, received the full brunt of Harrington's spleen in early July. A resolution had been passed refusing to employ or to deal with any person who was not a member of the National League. Describing the resolution as 'stupid and pernicious', Harrington fulminated:

> If your friends in O'Dorney could do nothing for the advancement of the national cause or could not make their branch efficient for really useful purposes it would be better for them shut the door altogether and not hold a mockery of a branch and pass idiotic resolutions calculated to injure the national cause.'[39]

According to Healy, there were two pre-requisites for a successful agitation; unity and adequate funds. In October 1885, Dillon had rashly promised that tenants entering into combinations would 'find the resources of the Irish National League at their backs'.[40] But when the parish priest of Farranfore applied for aid he was rebuffed by Harrington who suggested that the tenants should raise their own legal defence funds. 'If the fund comes from local resources', he concluded, 'the tenants themselves will be better able to realise the difficulties of the situation and to appreciate the importance of making a good fight'.[41] In December 1885, John Roche, the chairman of the Woodford Tenants Defence Association, received even shorter shrift from the National League. 'I was told, clearly and distinctly', he said, 'that the organisation would not then be responsible for anything in the shape of a combination'. He was told bluntly to advise the tenants to pay their rents.[42]

The refusal of the National League to recognise and, more importantly, to give financial assistance to striking tenants, precipitated the collapse of the majority of combinations on Irish estates, as Parnell had, no doubt, intended. He was at the apogee of his power and he dictated the policy of the National League through his control of the parliamentary party. In his quest for Home Rule, Parnell, quite simply , emasculated the agrarian agitation in Ireland.

It is impossible to assess accurately the extent of the rent strikes during the winter and spring of 1885-6. However, some indication may be gleaned from the extensive coverage given by the nationalist newspapers , especialy *United Ireland,* to the demands and actions of the tenants.[43] A more accurate pointer was a survey conducted by Arthur Mc Murrough Kavanagh, President of the Land Corporation of Ireland. Towards the end of 1885, he circularised some 800 landlords. In over 100 cases, covering 24 counties in the 4 provinces, he found that tenants 'had struck for a reduction'. Kavanagh confined his enquiries to estates on which judicial rents had been fixed.[44] Inevitably, the number of non-judicial tenants entering into combinations would have been considerably higher.

The rent strikes invariably collapsed when the landlords took legal action against the tenants. The Attorney-General, Hugh Holmes, in a conversation with the Earl of Carnarvon in November, said that tenants would pay when pressed, 'mainly because they dread to be put to the cost of legal payments'.[45] It soon transpired that, without the assistance of the National League, tenants were unwilling to take the extreme step of allowing themselves to be evicted nor were they prepared to have heavy legal costs heaped upon them. The Carlow and Wexford tenants of Arthur McMurrough Kavanagh had demanded a general abatement of twenty-five per cent. They banked their rents and started a defence fund. But when the landlord had writs served on twenty-one of the ringleaders, the combination quickly collapsed.[46] On a number of Kerry properties managed by George F. Trench there existed what he described as 'a solid strike... against the payment of rent'. The tenants had lodged their rents after their demand for a forty per cent reduction was refused. Trench served a number of writs and the tenants yielded on the day before evictions were due to commence, on condition that the legal costs were met by the agent.[47] Similar action was equally effective on the estate of Colonel James Crosbie of Ballyheigue Castle, County Kerry,[48] and on that of the Countess of Kingston at Mitchelstown, County Cork.[49] Some combinations showed greater resilience but tenants generally surrendered once the landlords commenced legal proceedings.

Despite their general lack of success, the rent strikes during the winter of 1885 reinforced the lessons of previous land agitations. The moral support of a national movement was necessary to encourage and unite a far from homogeneous peasantry. As the tenants were not altruistically motivated, the movement must be adequately financed. Tenants expected to be recompensed should legal proceedings be taken against them and to be supported if they were evicted. A crucial question was the disposal of legal costs. In the Ireland of the mid-1880s, agrarian distress was not a temporary phenomenon and, when the inevitable land agitation developed during the winter of 1886, the lessons of the previous year were applied. Tenant combinations during the winter of 1885 not alone contained the kernel of the Plan of Campaign but were a dress rehearsal for that agitation.

2 *Crisis*

> The government are face to face with an agrarian difficulty of grave and momentous consequence... . The question lies at the very roots of the social order for which Lord Salisbury evinces so much concern. *Freeman's Journal,* 14 August 1886.

On 7 June 1886, Gladstone's Home Rule Bill was defeated in the House of Commons by 341 votes to 311. Parliament was dissolved and, in the resulting election, the Unionists, comprising Tories and seceding Liberals, were returned with a majority of 118 over Gladstonian Liberals and Parnellites combined. Gladstone resigned on 20 July and Lord Salisbury again took office.

The refusal of the parliamentary party and the National League to countenance anti-rent combinations during the period of the Home Rule debates was greatly resented by many hard-pressed Irish farmers. To them, the prospect of Home Rule was secondary to the more pressing problem of rent and falling prices. Despite the efforts of the Parnellites, the countryside was seriously disturbed by widespread agrarian violence and outrages. In many areas, Captain Moonlight, battening on the prevailing agricultural distress, successfully filled the vacuum left by the failure of the parliamentarians to come to the tenants' aid.

At the Spring Assizes in 1886, the County Grand Juries were unanimous in demanding the restoration of law and order. Phrases like 'lawless agitation', 'social disorder' recur in their resolutions, while the word 'anarchy' crops up repeatedly. Kerry was the most disturbed Irish county, a point underlined by the resolution of its Grand Jury:

> Crimes of a serious character extensively prevail — murders, assaults with firearms, maiming of individuals, intimidation, boycotting, night attacks on dwellings, raids for arms, incendiary fires, mutilation and stealing of cattle, etc.... Unless the power of the law be speedily asserted and the parties whence these outrages emanated dissolved what remains of security and property will shortly be at an end.

The number of such crimes had almost doubled since the previous year, while the detection rate was a mere three per cent.[1]

Two separate correspondents to the *Times* testified in August 1886 that there had been little improvement in the condition of the south-west during the intervening six months. One referred to the 'dreadful state of persecution and terror' which 'the peaceable and loyal inhabitants' had endured for so long and warned that 'neither life nor property will be safe' during the coming winter.[2] The second wrote that Kerry was 'in a most extraordinary state. The whole county is in a condition of the most abject terrorism and the people are domineered over by a handful of village tyrants in each particular place'.[3]

Moonlighting was one symptom of tenant distress. Another was the continued existence of rent strikes on a number of estates in the poorer west and south-west, most notably in Kerry and Galway. According to John Roche, the Clanricarde tenants did not have the money to pay their rents and, despite the active discouragement of the National League, persisted with their combination simply because they had no other option.[4]

In January 1886, the tenants on a number of North Kerry properties, managed by George Sandes, demanded a thirty per cent rent reduction. When this was refused, Canon Davis of Listowel encouraged them to raise a defence fund and to act 'in union with one another'. The agent was boycotted after placards, displaying a crossbones and coffin motif, were posted throughout the Listowel district. When Sandes served writs and ejectment processes on a number of tenants, Canon Davis urged them to 'stow away their cattle'. Attitudes hardened during the summer and Davis played a major role in stiffening the tenants resistance. At the end of July he told them: 'You are the landlords' prop and their stay. Withdraw the prop, withdraw the people from the landlords and they fall to the ground like a tree felled by an axe'. In November, the agent was forced to admit that he had received virtually no rent during the previous year.[5]

In the Woodford area of County Galway, Lord Clanricarde, Sir Henry Burke and Mrs. Hannah Lewis had taken legal proceedings against their tenants early in 1886. The threat of evictions hanging over many peasants kept the agrarian cauldron on the boil. In March, Sir Henry Burke engaged the Property Defence Association to carry out evictions on his estate. Shortly after the arrival of emergency men in May, a crowd of between 500 and 800 assembled at Woodford. They formed into 'a funeral procession (and) marched twice up and down the town... preceeded by a coffin draped in black, on one side of which was an inscription "Down with Orangeism and Freemasonry" and on the other side "Down with Landlordism"'. Provocatively, the coffin was burned, 'amidst the wildest and most enthusiastic cheering of the crowd', near the spot where a process-server named Finlay had recently been murdered. The evictions on the Burke estate were abandoned after the tenants settled with their landlord at the beginning of June. However, the locality remained extremely disturbed and several evictions were carried out on both the Clanricarde and Lewis properties by the Property Defence Association in August.[6]

By early autumn 1886, agrarian agitation was once again widespread and could no longer be ignored. 'The land question presses forward for immediate settlement', warned the *Freeman's Journal* on 5 August, 'and it must be met'. *United Ireland* prophesied that the coming winter would inevitably see a struggle against 'tatterdemalion Irish spendthrifts', a 'battle between bankrupt landlordism for their rack rents and bankrupt tenants for existence fought out to the bitter end'.[7] Nationalist politicians and the Catholic hierarchy, who had championed Home Rule at the expense of the land question during the previous year, now concentrated on promoting the cause of the tenants.

The Irish parliamentary party, meeting in Dublin on 4 August, warned the government that the payment of judicial rents had become impossible and demanded the immediate revision of all such rents, the cancellation of arrears and the suspension of evictions. Additionally, the government should alter the rent-fixing clauses of the 1881 Land Act.[8]

On 17 August, Parnell arrived in London for the new parliamentary session. He informed John Morley that Irish rents could not be paid and that there would be trouble during the winter unless the government took some positive action. He believed that the parliamentary party should concentrate, for the immediate future, on land reform to help the tenants through the present crisis. In this way, Parnell hoped to prevent large-scale disturbance in Ireland.[9]

Archbishop Walsh, in an interview with T. P. Gill in the *Freeman's Journal,* said the land question was 'as gloomy as gloomy can be' and that 'sweeping reductions of rent' were absolutely necessary in many parts of the country.[10] At the close of their conference at Maynooth in early September, the Irish bishops expressed their anxiety at the inability of the tenants to pay the customary rents and at the increasing number of evictions. They feared that 'social evils of the gravest character' would result and urged the government to act.[11]

Despite nationalist foreboding, the English tory press perversely denied that conditions in Ireland were critical. The *Times* insisted that 'the signs of comfort are visible almost everywhere among the peasantry' and spurned all references to an agrarian crisis as 'arrant nonsense, born of ignorance and mischief making, if of nothing worse'. It rejected the claims that judicial rents could not be paid and concluded that in cases of genuine hardship landlords were giving liberal abatements.[12]

The implication was obvious. The government should reject all specious demands for ameliorative legislation and should concentrate on restoring law and order to Ireland. The Prime Minister had enthusiastically advanced similar sentiments on a number of recent occasions. On 15 May, at St. James' Hall, London, Salisbury outlined what John Morley described as his 'manacles and Manitoba' policy. He compared the Irish with Hottentots and said that neither was capable of self-government. He suggested that Irishmen should solve their economic woes by emigrating to Manitoba or else sufficient powers should be conferred on the government to enable it to administer Ireland properly. 'Apply that recipe honestly, consistently and resolutely for twenty years', he said, 'and at the end of that time you will find that Ireland will be fit to accept any laws that you may wish to give her'.[13] At the Lord Mayor's banquet on 11 August, Salisbury said that Ireland was 'not yet inhabited by a perfectly concordant and homogeneous people' and, in a reaffirmation of traditional Tory policy, emphasised that it was the government's duty and, by implication, intention to restore social order and to protect and support the loyalist population in Ireland.[14]

In the light of these statements, the opening proceedings of the new parliament were entirely predictable. There was no mention of any remedial legislation for Ireland in the Queen's speech. The government simply prevaricated by appointing parliamentary commissions to inquire into the Irish land system and Irish industry, a policy which was regarded by the nationalists as evasive and exasperating in view of the critical condition of Ireland.[15] Significantly, Salisbury announced that General Sir Redvers Buller was being sent to the south-west of Ireland to suppress crime and outrage.[16] The government was at pains to point out that the appointment was not a military one, nor was Buller being sent to establish martial law. His primary task was to reorganise the police force and its tactics in an attempt to restore law and order in Kerry and Clare. According to the Irish Chief Secretary, Sir Michael Hicks Beach, Buller was chosen because the government believed him to be experienced enough to carry out this task 'constitutionally, justly, humanely, uprightly and firmly'.[17] It was hoped that the Buller mission, if successful in Kerry and Clare, would serve as an example for the rest of Ireland, thereby increasing the morale and effectiveness of the police force and encouraging the Irish authorities to exhaust the ordinary law before applying for exceptional legislation.[18]

Buller's appointment offended liberals and nationalists alike. Morley sneered at it:

> To commission a general to hunt down little handfuls of peasants who, with blackened faces and rude firearms, crept stealthily in the dead of night round lonely cabins in the remote hillsides and glens of Kerry was hardly more sensible than it would be to send a squadron of life-guards to catch pickpockets in a London slum.[19]

Dillon believed that Buller, whom he termed the 'hero from Zululand', was sent to Kerry 'in the interests of the landlords . . . to put down by whatever force may be necessary every disorder which interfered with the exterminating policy being carried on there'. The end result, he maintained, would be counterproductive; by encouraging the Kerry landlords to carry on evicting, tenant resistance would intensify and the number of outrages would increase.[20] In general, the nationalist press took a similar view. The *Freeman's Journal* saw the appointment 'as a sustaining power for the landlords' and condemned what it described as the 'Prime Minister's Hottentot notions'.[21]

On the day that parliament reassembled and Salisbury announced that General Buller was being sent to Ireland, the Property Defence Association began to evict tenants from the Clanricarde estate. For more than a week, the spotlight fell on Woodford as emergency men and bailiffs, protected by a force of some 500 policemen, tried to evict a small number of tenants. The latter barricaded their houses and tore up access roads; trees were felled and bridges pulled down. 'Saunder's Fort' was defended heroically. Sticks, stones, boat-hooks and gaffs, as well as swarms of bees, lime and boiling water were used by the tenants to repel the repeated assaults of the bailiffs. At the end of this extraordinary campaign, four evictions had been effected and twenty-one young men arrested and sentenced to prison terms ranging from twelve to eighteen months.[22]

The Woodford area, which had been seriously disturbed since the previous winter, was inflamed by the evictions on the Clanricarde estate and the subsequent ones on the neighbouring property of Hannah Lewis. The farmers of south Galway were urged to assemble in their thousands at Portumna on Sunday 26 September 'to denounce the action of Lord Clan-Rack-Rent and Blind Lewis, the Priest Prosecutor', to 'protest against the hydra-headed monster, landlordism' and to 'show the Orange emergency men that we are prepared for the battle'. The notices advertising the meeting, which were extensively posted throughout County Galway, concluded: 'We will fight the tyrants in a constitutional manner and teach them that the people of Ireland are a stronger element than cruel and oppressive landlordism'.[23]

Parnell warned the members of the House of Commons that there existed in Ireland 'a state of affairs more intensified in their probable evil effects than existed even in 1880' and that, unless remedial legislation was introduced, neither he nor the Irish members would be able to control the passions of the tenants during the coming winter.[24] He proposed a Tenants Relief Bill and, in moving its second reading on 20 September, said the tenants could not be left to the mercy of the landlords. 'The history of Irish landlords is a history of which they cannot be proud... these men whom you have put in as stewards have been false to their trust and have oppressed, ill-treated and rack-rented their tenants'. The bill contained three clauses: the Land Commission should be empowered to reduce judicial rents fixed before 31 December 1884 on condition that the tenants paid half the rent due in the current year and half their previous arrears; leaseholders should be admitted to the benefits of the Land Act of 1881; and the courts should be empowered to suspend all eviction proceedings on payment of half the rent and arrears.[25] Dillon described the bill as 'moderate'. It was introduced, he wrote, to compel all Irish landlords to act reasonably and humanely towards their tenants, thus ensuring 'peace and quiet' in Ireland during the winter.[26] Both he and Parnell were anxious to cement the Liberal alliance and were apprehensive lest it be undermined by trouble in Ireland.[27]

The defeat of the bill was inevitable. Any tampering with judicial rents, regarded as sacrosanct by the Tories, was bound to be stoutly resisted. The Irish Chief Secretary expressed the general feeling of the party when he described the bill as 'an act of gross injustice to the landlords of Ireland'.[28] He confided to Buller: 'Parnell's bill is a coercion bill for all landlords...It is too extreme a measure of coercion to be accepted by us'.[29] The rejection of the Tenants Relief Bill proved to be a turning point. The government, by its insensitive handling of Irish problems, was largely responsible for making a new and organised land struggle inevitable. It consistently and stubbornly refused to acknowledge Irish distress. As the economic depression approached its nadir, the executive postponed the possibility of remedial legislation, for a number of months at least, by appointing parliamentary commissions of inquiry. Salisbury seemed determined to restore law and order while ignoring the underlying cause of that disorder. The appointment of Buller further alienated the nationalist population and convinced them that the government was intent on supporting Irish landowners at the expense of the tenants. The rejection of Parnell's bill convinced Dillon and others of the futility of relying on the legislature. The tenants could expect little sympathy, it was acknowledged, from a Tory-controlled House of Commons and would have to rely on their own resources and to resort, for remedies, to the tactics of the Land League days. Dillon announced indignantly in the House of Commons:

> I go back to Ireland, to tell the tenant that if he wishes to live he must trust to his own exertions, and not to this House; and so long as I have life and liberty...I shall tell the people of Ireland to continue in that course of persistent and determined agitation by which in the past they have won every single liberty and every single concession which has been granted them, and by which in the future, if they only show perseverance and bravery, they will win in spite of the Irish landlords and in spite of this House, the right to live as freemen in the land of their birth.[30]

On 28 September, the National League publicly intimated its support for a new land struggle. The central committee acknowledged that large numbers of tenants would have to turn to them for support during the winter. It had already been arranged, according to Harrington, that all funds contributed to the League in Ireland would be devoted exclusively to the evicted tenants.[31] Dillon pledged that the National League would do everything in its power to assist the people[32] and Parnell wrote to the president of the National League of America, asking support for an anti-eviction fund.[33] The Irish leaders, while acknowledging that a new land agitation was inevitable, gave no indication as to what course it might take. No definite policy had been formulated nor did they have any specific advice to offer the tenants beyond the trite promptings of Dillon that they should organise on individual estates, should 'stick together' and 'make just and reasonable demands'.[34] Within three weeks, however, Dillon was outlining the daring and ingenious scheme that was to dominate Irish political and social life for the remainder of the decade.

3 *Ultima Ratio Regum*

Jim from Castleisland meeting Mick from Glenbeigh asks, "Will Mick an' how are ye getting on?" "Illigant, glory be to the saints." "How's that, Mick? Sure prices is low." "Thrue for you Jim, prices *is* low, but what we *has* we *has*, for we pays nobody."

Samuel Murray Hussey, *Murray's Magazine*, Vol. 1, May 1887.

(i)

Constable J. J. Noble was the police reporter at the Woodford meeting on 17 October 1886 at which Dillon unveiled the new policy of the National League. Noble noted the tension and excitement that prevailed among the 1,000 assembled tenants. Bands played patriotic airs; priests, members of parliament and William O'Brien delivered exhortatory speeches. At the outset, the parish priest of Woodford, the Rev. P. Coen, requested the constabulary to withdraw, from their position in front of the platform, to the outskirts of the crowd. This resulted in a 'tremendous outburst of boohing and groaning' and the police were bundled back to the periphery. 'During the whole time the police were being cursed and shoved by the crowd', Constable Noble observed, 'a smile of satisfaction and approval overspread the features of Rev. Fathers Egan and Coen and other gentlemen on the platform while an angry frown was clearly visible on those of Messrs. Dillon and O'Brien.[1]

The new policy expounded by Dillon was clear and unambiguous and may be summarised as a device for collective bargaining on individual estates.[2] A week later, the Plan of Campaign was first published in *United Ireland*. It was essentially a more comprehensive and specific treatment of the advice given by Dillon to the Woodford tenants or, as *United Ireland* phrased it: 'It harmonises with the plan outlined by Mr Dillon at Woodford last Sunday, but it fills in the outlines and it dots the i's'.[3]

The Plan of Campaign manifesto can be divided into four sections. In a short preamble, the author states that, as the tenants were unable

Supplement Gratis with
"UNITED IRELAND."
Saturday, Oct. 23rd, 1886.
COERCION & NO RENT
SURRENDER AT DISCRETION
LANDLORDISM ON THE HORNS OF A DILEMMA.

to pay the present rents, an increase in evictions was inevitable and was bound to lead to an agrarian agitation during the winter. The core of the document advised the tenants on the course they should follow. On each estate, they should meet to decide on the extent of the rent reduction they wanted. Should that abatement be refused, the proferred rent was to be handed over to a managing committee who would lodge the so-called 'estate fund' with a trusted individual known only to themselves. This money was to be used to fight the landlord and support the evicted. It was strongly emphasised that under no circumstances should tenants pay legal costs and, during the next few years, this was the one principle on which the leaders of the agitation refused to compromise. In the unlikely event of a dishonest trustee, the National League would make good the embezzled funds. Furthermore, the League promised to support those evicted should the estate fund become exhausted. Penultimately, the legal remedies available to landlords against defaulting tenants were detailed. Finally, the manifesto recommended that the fullest publicity be given to evictions and urged that those responsible be rigorously boycotted.[4]

(ii)

For legal reasons the Plan of Campaign was published anonymously but its author was Timothy Harrington, a lawyer and secretary of the National League.[5] It was drawn up at a meeting of the organising committee of the League sometime after the rejection of Parnell's Tenants Relief Bill, probably during the first half of October 1886.[6] A suggestion put forward by William O'Brien and amended by Harrington was the chrysalis from which the Plan of Campaign developed.[7] Harrington's amendment had been suggested to him by Healy from a chance remark made by Parnell after the failure of the 'No Rent Manifesto' in 1882. The latter had then vowed that he would never again lead an agrarian movement unless the tenants lodged seventy five per cent of their rents in a common fund.[8] It is ironic, given his attitude to the Plan of Campaign, that its *modus operandi* should have been unwittingly suggested by Parnell.

Ingeniously, the opponents of the Plan traced its inspiration to the writings of James Fintan Lalor, who had advised the Irish farmers to refuse to pay rents or arrears in the autumn of 1848 until they had adequately provided for themselves and their families for the ensuing twelve months. Lalor suggested that the claims of the tenants, 'the true proprietors' had precedence over those of the landlords, 'the present usurping proprietors'. The people should decide at a national convention how much rent they would pay and to whom.[9] Essentially, the Plan of Campaign rested on the same principle: the tenants fixed their own rents.

The Plan was shaped by the observations and suggestions of several individuals, including Lalor, but if any single person is to be credited with inspiring the scheme it is Tim Healy. In the autumn of 1885, Healy had advised the tenants to lodge a reduced rent with trustees. Should the landlord refuse to abate the rents, the money deposited was to be used for the tenants' benefit.[10] There is a striking similarity between Healy's speech on 30 August 1885, in which he outlined his ideas, and that delivered by Dillon at Woodford on 17 October 1886. Indeed, one nationalist newspaper reported an address by Healy in October 1885 under the heading, 'A Plan of Campaign'.[11]

Dillon and O'Brien guided the fortunes of the movement during its five-year existence with extraordinary energy, courage and selflessness. O'Brien had the more attractive personality and was, unquestionably, the most popular nationalist leader of the day, but his very exuberance and lack of caution predisposed him to making enormous political gaffes, such as his quixotic trip to Canada in 1887 to highlight evictions on the Queen's County estate of the Marquess of Lansdowne, Governor-General of Canada, and his reckless encouragement of 'New Tipperary' in 1889. Dillon was acknowledged as the leader of the agitation and was generally deferred to by O'Brien. 'Honest' John Dillon was respected more than liked by his political allies and was loathed by his opponents. A week after the publication of the Plan of Campaign, the landlord Dublin *Daily Express* described him as

> a fire-eating anarchist... a stirrer-up of strife and ill-feeling, a man devoting his energies and talents to the vile business of stirring up class against class and sowing wherever he goes the seeds of contention and confusion. If there be bad work in Ireland, it will have originated mainly in the action and language of this smooth-spoken gentleman.[12]

After Dillon and O'Brien, the most important campaigner was David Sheehy, who boasted in November 1887 that he had promoted the agitation on thirty-five different properties.[13] In the following May, he invoked 'God's blessing on the Plan of Campaign' and vowed to uphold its banner until the fight was won.[14] About half the members of the parliamentary party associated themselves, with varying degrees of fervour, with the agitation. Willie Redmond,

Matt Harris, Daniel Crilly, Luke Hayden, W. J. Lane, Tom Condon, John O'Connor, J. R. Cox, Dr Tanner and, to a lesser extent, John Redmond, Tim Healy and T. P. Gill, were the most important. Tim Harrington drafted the Plan of Campaign manifesto but beyond that his role was negligible. Micheal Davitt was in America during the formative months of the agitation and, though he addressed a few agrarian meetings on his return, he frequently censured the Plan on the grounds that it was too moderate.

> 'The fault I found with the Plan of Campaign', he complained on New Year's Day 1888, 'was that it was not half advanced enough for me. It was only passed as a temporary expedient. It was only worked for the immediate relief of the tenantry of Ireland, whereas what Ireland wants, what we want and what justice wants is the life of Irish landlordism'.[15]

Whether wilfully or otherwise, he ignored the reality of the agitation. Dillon, no less than Davitt, was an unwavering opponent of the institution of landlordism which he described as 'the curse of Ireland' and he frequently denounced Irish landlords as parasites.[16] Nowhere is Dillon's hatred of the system and the implication of the Plan of Campaign more clearly seen than in the powerful philippic he delivered at Castlerea on 5 December 1886:

> Now is the hour to destroy the tyranny of landlordism and assert the freedom of the Irish tenants. I say that this struggle in which we are engaged is as holy a struggle as any nation in the world ever went into. It is a struggle to undo the system set up in this country by William and Cromwell and under which the Irish people have toiled for nearly 200 years. It is a struggle to make the Irish farmer a free man in his home instead of being what he has been for nearly 200 years, a serf living from day to day in fear of a master who cared little for Ireland and less for him. It is a struggle to hand over the possession of the soil of Ireland to the class who labour in that soil, who love their country and whose fathers have fought for it; to make them free and to emancipate them from the rule of men who swept away from the Irish farmer nearly all the product of his industry to another country, who reproach us with our poverty, insult us and tell us that our people can't live except on charity. We call on the people this winter . . . to fight for free farms and freedom and to fight in order to destroy a system which has eaten into the manhood of our race like a corroding ulcer.[17]

O'Brien, Harris, Sheehy, Crilly and the brothers Redmond publicly condemned landlordism and shared O'Brien's belief that the Plan of Campaign would settle 'the land question and the peasant proprietary question and the Home Rule question at one blow'.[18] Any remaining doubts about the purport of the agitation were graphically and explicitly dispelled by Harris at Castlerea:

> This is not a question between small reductions and large reductions. It is not a question of what a man is able to pay at present prices. And it is not a question of good landlords and bad landlords... The great mighty question is whether the lands of Ireland shall belong to the people of Ireland or whether they shall belong to the enemies of the people. I am not going to indulge in a 'No Rent Manifesto' but we put a programme before you that will lead to that result—that will first take one slice, then take a second slice and we will keep slicing at it until nothing remains.[19]

The Plan of Campaign was a reaction to the economic depression of the mid-1880s and, in particular, the failure of the Tory government to ameliorate the lot of the Irish tenants in the autumn of 1886. Although it was rooted in economic difficulties, the agitation was always more than just a reaction to negative economic circumstances. Promoters of the Plan stressed that it was a temporary expedient to be used solely against harsh or exacting landlords. However, no attempt was made to define such individuals while the Plan of Campaign manifesto merely instructed the tenants on 'how to meet the November demand'. It was left to the tenants on the different estates to determine whether or not their landlords were rack-renters. Inevitably, criteria other than tenant necessity came to apply. The Plan had an implicit political role almost from the beginning and it soon became apparent that, for the leading advocates of the agitation, its purpose was no less than the extirpation of landlordism. It was not alone an attack on the institution of landlordism but, in the wake of the rejection of the Home Rule Bill, it was for many an attack on the landlord class as part of the English garrison in Ireland, underlining once again the close relationship between land and nationalism.

This is not altogether surprising. The leading campaigners came from the more radical wing of the parliamentary party and several had strong connections with Fenianism. According to the Special Branch, David Sheehy had been a high-ranking member of the I.R.B. but had resigned when he was elected to parliament. At a National League meeting in October 1888, J.C. Flynn openly boasted of his connection with the physical force movement. 'He was proud of having belonged to it', he said. Tom Condon, a

Clonmel butcher who was known as 'The Tipperary Dragoon', was an active Fenian. After being selected as a parliamentary candidate for Tipperary in November 1885, he claimed that he was 'no shilly-shally milk and water politician' but was proud to follow in the footsteps of Mitchel and Rossa. Daniel Crilly and Joseph R. Cox, who were active in promoting the Plan in Mayo and Roscommon, were prominent members of the I.R.B. During the autumn of 1887, Cox apparently used the Plan of Campaign as a cover to recruit for the Fenians.[20]

John Kelly and John Cullinane were paid organisers of the Plan. Kelly, who was related to Tim Harrington, promoted the agitation on Lord Massereene's estate in Louth and on those of Mr Brooke in Wexford and of Wybrants Olphert in Donegal.[21] Cullinane was born in Bansha, County Tipperary, in 1856. When he left school, he became a paid Centre in the Fenian organisation. He closely identified himself with the Plan of Campaign struggles on the Kingston and Ponsonby Properties in County Cork. He played a prominent role in the dispute with A. H. Smith-Barry which commenced in Tipperary in mid-1889. The District Inspector's report on Cullinane concluded:

> He has paid no rent for six years. He lives in a good hotel in Tipperary, drives about the county and superintends the building of New Tipperary. He is a member of the Tenants Defence Association committee and gets at present twenty pounds per month... But while actively engaged in illegal work he carefully avoids any words or acts which could be used in evidence against him by the police.'[22]

The local leadership of the Plan of Campaign was drawn almost exclusively from the ranks of the Catholic clergy, although a few lay leaders of note emerged during the course of the agitation, such as John Roche, John Sweeney and Francis Tully from County Galway, John Mandeville, a gentleman-farmer from Mitchelstown, and John Fitzgibbon, a Castlerea draper. Mandeville was a friend and associate of William O'Brien and was widely respected throughout County Cork. Fitzgibbon 'played a major role in spreading and maintaining the Plan throughout County Roscommon'.[23] He told an assembly of farmers in April 1888 that, as president of the Castlerea branch of the National League, 'the peace and the keeping of law and order in this part of the countryside has, to a certain extent, been vested in my hands'.[24] Roche was associated with the struggles on the Clanricarde and Lewis estates in Galway. 'We have a tyrant to crush today', he said of Lewis in December 1886, 'throttle him and don't let him go 'till you have knocked the glass eye out of his head'. [25] Sweeney, from Loughrea, was a high-ranking member of the I.R.B. and played a prominent role in the Clanricarde, Dunsandle and Millar disputes.[26] Tully, from Looscaun, was affectionately known as 'the doctor' because of his penchant for prescribing 'leaden pills' for landlords, agents and bailiffs alike. Between November 1886 and July 1890, he was prosecuted no less than six times for assault, intimidation and unlawful assembly and was imprisoned on a number of occasions. He too was a member of the I.R.B.[27]

(iii)

A visitor to Ireland in the winter of 1887–8, who preferred to remain anonymous, noted that the vast majority of the clergy had a small farming background or were the sons of shopkeepers. They are 'the very flesh of the flesh of the people', he wrote. 'In appearance, feature, pronunciation, careless grammar, if not vulgarity of phrase, they are . . . a peasant priesthood . . . The Catholic clergy are part and parcel of the people'.[28] They were alive to the wants of the people and closely identified with their feelings and aspirations. Apart from a minority of 'Castle' bishops and priests, the clergy shared the attitude of the nationalist community at large towards the landlords. At the end of October 1886, the Loughrea curate, John W. Fahy, who was at the time languishing in Galway gaol, described landlordism as an 'iniquitous system . . . which has kept the country in a state of chronic disorder and turmoil, retarded the progress of commercial activity in our provincial towns, destroyed the growth of industrial and agricultural prosperity and hitherto poisoned the sources which lead to national greatness and happiness'.[29] In September 1887, a County Armagh priest urged Lord Londonderry to follow his ancestor's example and cut his own throat.[30] Fr Lane of Ballinadee, County Cork, pledged to continue the agitation until landlordism was 'utterly crushed'.[31] At Kells, County Meath, Fr Woods promised that the clergy would teach the people 'to conspire, confederate and combine'. They would plant the banner of the Plan of Campaign 'on the demolished ruins of landlordism, the British government and injustice in the country', he said.[32] Fr Gaughran of Carrickmacross vowed to 'trample upon landlordism with the help of God'.[33] While he did not invoke supernatural aid, a County Louth priest prescribed a similar fate for the landlords, whom he claimed were on the last ditch. He urged the people to push them off 'and when down tumble the ditch down on them and then jump on them'.[34] Nowhere is the attitude of priests, people and politicians, to landlordism and the Plan of Campaign, more unambiguously and unapologetically expressed than in a speech delivered by Fr Casey, the parish priest of Mountcollins, County Limerick, on St Patrick's Day 1888:

> I don't care who the landlord is, no matter what sort he is, if he was an angel from heaven he is a bad man provided he is a landlord. Well, some of you may say we have a little bit of cheap land here. I can tell you it is not a bit too cheap at all. It is worth nothing at all whatever and if you get a thousand acres of it for half a farthing I say join the Plan of Campaign and ask a reduction of fifty per cent.[35]

Several priests secured national recognition, some notoriety and a few prison sentences for their role in the agitation. The most important were: Donal Keller, the parish priest of Youghal; David Humphreys of Tipperary; Father Kennedy of Meelin, County Cork (who described the local landlord, W. N. Leader, as 'a penniless and destitute exterminator, a fangless boar');[36] Matthew Ryan of Herbertstown, County Limerick, who was popularly known as 'the general' and believed that no rent was a fair rent;[37] and James McFadden, parish priest of Gweedore, County Donegal.

McFadden was a Messianic figure whom the people regarded with 'a reverence and awe scarcely credible', noted Allan Cameron, the Divisional Commissioner.[38] The police described him as 'a red hot nationalist partisan, an ambitious, unscrupulous, clever agitator of the most dangerous type'.[39] McFadden unapologetically considered himself to be 'the law in Gweedore'.[40]

With the aid of his acolyte, Father Daniel Stephens, McFadden introduced the Plan of Campaign to five County Donegal estates.[41] 'Any man of you who can spare a year's rent will put it in the Plan of Campaign', he instructed an assembly of tenants in September 1887. 'You will put it in minus one third. You will put it in to one of those strong iron boxes. Let us not go back now we are in the midst of the battle'.[42]

In November 1887, Captain Hill of Gweedore surrendered, on penal terms, to the Plan. He conceded an all-round reduction of 30 per cent, reinstated some 130 evicted tenants, paid £900 legal costs and reduced the tenants' arrears of £3,400 by 60 per cent.[43] A jubilant McFadden pledged to continue the struggle against landlordism. 'So long as a landlord owns the breadth of a shoe of land in Gweedore', he said on 4 December 1887, 'so long will we fight against them'.[44]

He now turned his attention to his 'old personal antagonist', Wybrants Olphert of Falcarragh.[45]

In early 1888, McFadden and Stephens were prosecuted under the Criminal Law and Procedure Act of 1887 for advocating the Plan of Campaign.[46] On 18 January, a rumour spread throughout the village of Crossroads that Stephens was to be arrested. At noon on the following day, reported Sergeant Patrick McPhillips, 'a horn was blown towards Muckish Gap and answered by another towards Gortahork and immediately people began to stream into the village from all directions, carrying bludgeons. They formed a procession with three green banners and marched through the village, halting before the priest's lodgings. Nearly 900 persons were present'.

Stephens told them that it was the proudest day of his life and that he was quite prepared to go to prison. 'I am willing to make the sacrifice', he explained, 'because it is in the cause of the poor and the oppressed. I would enter prison as joyfully as I would go to a wedding'. He reasoned, rather tortuously, that imprisonment would give him, an obscure country curate, a certain notoriety and influence with the leaders of the movement in Dublin. An appeal for funds would produce 'a stream of gold' which would enable them to continue the struggle until they got 'a fair, reasonable and straightforward offer'.[47]

On the following day, 20 January, McFadden was arrested in Armagh, where he had gone to commemorate the recently deceased Archbishop, Daniel McGettigan. The Donegal people were incensed, not so much by the arrest itself, which was not entirely unexpected, as by its manner. Early on the following morning, 400 men from Gweedore, accompanied by a fife and drum band, arrived in Crossroads en route for Dunfanaghy where they expected to intercept McFadden on his way to Derry. By early afternoon, the gathering had swelled to 1,000 men. They stood in front of the police barracks and 'cheered, booed and waved bludgeons with which the Gweedore contingent were all armed'. Father McNelis protested at the apprehension of Father McFadden 'by men who would be afraid to come to Gweedore to arrest him'. Father Stephens deprecated the treatment meted out to their 'beloved and patriot priest'. 'These cowards', he accused, 'dare not come and touch him among his own people because they know that not all the forces at their command, aye, all the Queen's army, could take him from his people in Gweedore'.

Later that evening, Sergeant McPhillips reported:

A tar barrel was lighted in a back yard and shot out on a cart to the

> street and immediately after all the nationalists' windows (and also Brewster's Hotel) were illuminated with candles placed inside almost every pane in honour of Mr O'Brien's release.[48] The people of the village turned out with a drum and a melodeon and were soon augmented by arrivals from the surrounding districts to the number of about 300. They marched to and fro through the village and twice out the principal roads leading from it, playing (and some girls singing) "God Save Ireland", "O'Donnell Aboo", "St. Patrick's Day", etc. They frequently halted before a house in which Father Stephens was visiting and once before his lodgings when he returned and cheered for him, for Father McFadden of Gweedore, the Plan of Campaign, O'Brien, Davitt, Parnell and gave groans for Mr. Balfour, the landlords and the police.[49]

McFadden's case came up for trial at Dunfanaghy on 29 January 1888. The police reported that 'an enormous crowd of people assembled with bands, banners, etc. and Father McFadden had to be escorted by a troop of dragoons and a strong body of constabulary'.[50] Archbishop Logue noted that about twenty priests kept the twelve thousand assembled tenants under perfect control. 'A military officer present said that he never saw such an immense multitude so quiet before. He said the people seemed more afraid of the umbrellas of the priests than of the rifles of his men'.[51] McFadden was sentenced to three months imprisonment, a decision he appealed to Letterkenny Quarter Sessions on 19 April, and was released on bail.

On 17 April, while McFadden and Stephens were on their way to Letterkenny, a demonstration in their honour was held at Termon. Their only crime, said the local curate, was that as priests they had stood between the people and their oppressors. Stephens replied that it was 'an honour and glory to suffer in the cause of our grand old country'. An uncompromising McFadden stated that they had entered into 'a solemn compact, made beneath the canopy of heaven in the presence of God', that they would not settle with the landlords until their terms had been granted. 'We adopted the Plan of Campaign as a matter of necessity', he concluded, 'we were dealing with a hard-hearted cruel lot of landlords'. At a similar demonstration at Cresslough, Stephens warned that the government had made a grave mistake in laying 'unholy hands upon the Irish priests', while McFadden called on the tenants of Ireland to refuse to pay any rent 'until the landlords make restitution for the outrages they have committed and made restitution to the Irish priests for what they have suffered'.[52]

McFadden's sentence was increased to six months on appeal. On his release on 20 October 1888, he was unrepentant and boasted that he was unchanged and unchangeable. 'I find my enemies almost grovelling at my knees tonight', he exulted, 'and I am the victor'. He promised to continue the agitation until he had planted the flag of victory on the grounds of Donegal.[53]

Evictions which had been pending on the Olphert estate for some time were scheduled to begin on 2 January 1889. Stephens urged McFadden to attend:

> Wherever then you maybe, whatever engagements you have on hands, lay all aside to rush to the scene of combat. You know what Olphert is. You know it will take all our united energies to beat him down. You know you have often wished for an opportunity to get the people united in stout battle against him. That opportunity is at hand. Do not, I beg of you, let it pass... Our success depends on you being present.[54]

The evictions lasted for a week and were attended by 'terrible scenes of violence'. According to the police, the resistance which the tenants offered to police and bailiffs alike was deliberately orchestrated by Stephens and McFadden.[55] The latter was summoned to appear before Dunbeg court on 28 January and when he failed to appear a warrant was issued for his arrest. When the police attempted to apprehend him outside his church on Sunday 3 February, a mêlée ensued in which District Inspector Martin lost his life in very dramatic circumstances. The priest and a large number of his parishioners were arraigned for murder. The government considered it impolitic to proceed against McFadden on a capital charge. Instead he was blackmailed into pressurising his fellow-prisoners to plead guilty to manslaughter. At their trial in Maryborough in October 1889, the crown dropped the capital charge against McFadden. He pleaded guilty to obstructing the police and was allowed to go free while his co-defendants were sentenced to long terms of imprisonment for manslaughter.[56]

In November, a jubilant Divisional Commissioner reported:

> The collapse of the defence of the Maryborough prisoners and the desertion of these poor dupes by Fr McFadden in the hour of their sorest need has been such a blow to the power he so arrogantly assumed that he will never really recover; and it will be hard for him again to assert with any prospect of being believed by his own ignorant, credulous people that he is *in propria persona*, "the law in Gweedore".[57]

McFadden was forbidden by his bishop to involve himself further in political agitation and was, shortly afterwards, transferred to another parish.[58]

A minority among the Catholic hierarchy strongly disapproved of the Plan of Campaign, notably Dr O'Dwyer, Bishop of Limerick, described by Wilfrid Blunt as a 'shaky kind of patriot' who had gone over bag and baggage to the enemy, and Dr Healy, Coadjutor Bishop of Clonfert, whom Blunt pilloried as 'a lover of purple and fine linen . . . a special friend of the landlords'.[59] In an emotional and self-righteous response to a charge in the *Freeman's Journal*, O'Dwyer denied that he was a unionist or a 'landlord' bishop and proceeded to condemn boycotting and the Plan of Campaign, one because it was sinful and the other on account of its manifest unjustness. He felt conscience-bound, he wrote, to stand aloof from a movement which he castigated as politically stupid and morally wrong.[60] Healy telegraphed his agreement with every word expressed by O'Dwyer.[61] The Bishop of Kerry 'in no way sympathised with or favoured the movement',[62] while the Bishop of Waterford opposed it 'as bad means in a good cause'.[63]

The vast majority of the bishops, however, supported the Plan of Campaign, albeit with varying degrees of enthusiasm. Some feared the pull of the secret societies if the episcopacy refused to align themselves with the popular agitation.[64] Others considered the Plan to be a desperate remedy for a desperate situation.[65] If we are to believe the bigoted and anti-clerical Colonel Turner, the Divisional Commissioner for the south-west, the Bishop of Killaloe, on account of his 'age and imbecility', cared little one way or the other. As a consequence, exploded Turner, the Clare priests 'run riot in the paths of wickedness' and 'incite the people to break the law, to boycott and intimidate and join the vile conspiracy called the Plan of Campaign'.[66] Most of the bishops followed the example of the two most influential churchmen of the day, the formidable Archbishop of Dublin, Dr Walsh, and his ebullient colleague in Cashel, Dr Croke, who, after initial doubts, gave their unequivocal support to the agitation. In early December 1886, Walsh publicly defended the Plan of Campaign on moral grounds.[67] The result of his archiepiscopal blessing was to convert the waverers in the nationalist ranks. As the Bishop of Clonfert wrote to his nephew in Australia a fortnight later, '*practically* the priests, bishops and people are united'.[68]

Dr Duggan was referring, albeit unconsciously, to the Catholic populace; Irish Protestants, as a rule, dissociated themselves from both the National League and the Plan of Campaign. Fr John Maher, who presided at a meeting of the Ballyadams and Wolfhill National League in mid February 1887, expressed his sorrow at the refusal of local Protestants to join forces, politically, with their Catholic neighbours.[69] A week later, Fr J. McGuire, chairman of the Moneygall, Barna and Dunkerrin National League, made a powerful appeal for the political support of the Protestant shopkeepers and tenant farmers of Moneygall. He urged them to lay aside sectarian feeling which, he said, was the bane of Ireland.[70] According to Wilfrid Blunt, the records of the National League disclosed that its membership was almost exclusively Catholic and that the organisation was dominated at local level by the parochial clergy.[71] Invariably, the grass-roots response to the Plan of Campaign was orchestrated by the same individuals. In May 1889, the Irish Chief Secretary, Arthur Balfour, informed the Nonconformist Unionist Association in London that

> it had been a matter of profound surprise to those who knew little of the working of the Roman Catholic Church in Ireland that priests of that community preached boycotting and a parish priest or curate was always to be found implicated in every case where the Plan had been started.

Balfour concluded that Protestant tenants had, as a rule, 'refrained from joining in those illegal conspiracies'.[72] Protestant farmers on the estates of Lord Massereene in Meath and Louth, Lord Lansdowne in Queen's County, George F. Brooke in Wexford and H. J. Murray-Stewart in Donegal refused to join their Catholic fellow tenants in the Plan of Campaign. Protestant tenants on Wybrant Olphert's Donegal property constituted 'a mere handful' but, according to the Liberal-Unionist MP for South Tyrone, 'they have steadily resisted the Plan of Campaign and striven to be honest men'.[73]

(iv)

On successive Sundays during the winter of 1886 and throughout 1887, nationalist MPs stumped the country, advocating and lauding the Plan of Campaign. Eight hundred tenants attended at Kilbrin, County Cork on 19 December 1886 to mark ' the opening of the glorious campaign at Duhallow'.[74] Three thousand Kerry farmers thronged into Dingle in mid-January 1887 to hear Edward Harrington MP and Charles Coneybeare, MP for Cornwall, propound the new agitation. The visitors were escorted into the town by '300 horsemen bedecked with green sashes and ribbons'.[75] When Dillon and O'Brien arrived at Portumna on 18 December 1886, the horses were taken out of the shafts and the car was dragged through the streets of the town, while the local brass band played, 'See the conquering hero comes'.[76]

The nationwide public meetings were popular manifestations of nationalist fervour. Crucially, they promoted solidarity among the tenants and fostered the burgeoning agitation. Priests and politicians alike delivered exhortatory speeches, the complexities, as well as the politics and tactics, of the Plan were expounded and resolutions in its favour were adopted. The police acknowledged the impact of the weekly meetings:

> The people showed considerable excitement under the influence of the promises held out to them of a brilliant future and the denunciation of the government, police, landlords and bailiffs. The language used by the speakers at these meetings was very strong and was full of expressions of hostility to government and law. The public meetings which have been held have caused a terrible change for the worse in the feelings of the people and have done great harm.[77]

The stump orators employed a variety of arguments in order to encourage the tenant farmers to adopt the Plan of Campaign. O'Brien's claim, that the initiative lay with the tenants and that they would be 'fools and jackasses, as well as traitors', to pay without having the judicial rents reduced by at least twenty five per cent, had a powerful appeal.[78] 'Will you be bosthoons', asked David Sheehy rhetorically, 'to go on paying rack rents, paying impossible rents, starving your families, robbing your shopkeeper or cheating the banks?'.[79] O'Brien promised Lord Lansdowne's tenants that they would come out of the struggle 'richer and stronger men . . . richer in purse, richer in property, richer in honour and in the love and the eternal gratitude of their fellow countrymen'.[80]

Tenant altruism and patriotism were exploited by nationalist politicians and newspapers alike. 'They were not only fighting for bread and butter and the right to live', proselytised Willie Redmond, 'they were also fighting in a deathless struggle and a holy cause, that of the nationhood of Ireland and 'though the clouds were thick over their heads now the sun of freedom would yet shine on the happy homes of a free people and the green banner of their dear native land'.[81] If the Clanricarde tenants adopted the Plan of Campaign, Dillon stressed, they would not alone encourage other tenants but would serve 'the cause of Ireland and of Irish nationality'.[82] Daniel Crilly told Lord

Massereene's tenants that they should be prepared to undergo imprisonment to further the agitation. 'If they have not the courage', he opined, 'they are traitors to the cause of Ireland'.[83]

Tenants who successfully resisted the blandishments of the politicians were unable to withstand the intimidation that was directed against them. Fear may have been a base motive for upholding the principles of the agitation but, nonetheless, it was a compelling one. The military analogy, as well as militaristic language, was a characteristic of the Plan of Campaign years. It was expounded at its most chilling by Dillon in April 1888, when he delivered the following warning to the O'Grady tenants:

> I have always said that going into this struggle is very like going into a battle and in a battle when a man turns tail his officer will shoot him through the head. You know perfectly well what I mean. I don't say we should use personal violence but it is a principle of the army or any body of disciplined men that when people go in voluntarily into a tremendous difficulty the man who turns tail on the rest is looked on as a traitor.[84]

On 6 January 1888, James McLoughlin, a tenant on the Worthington estate in County Roscommon, was attacked while on his way to Ballingar fair by six men with blackened faces. He was badly beaten and his pigs were scarred, rendering them unsaleable. McLoughlin had refused to join the Plan of Campaign, as he believed the terms offered by his landlord were reasonable.[85] Owen Keenan was the only tenant on the Packenham-Mahon property in Roscommon who refused to join the Plan. He paid his rent and was promptly boycotted. Keenan believed his life was in danger after he found a party of twenty men near his house. He was afraid to stay at home and spent six successive nights in the bog. Finally, out of sheer terror, 'he sold two pigs at a loss and paid his rent into the Plan of Campaign'.[86] Animals belonging to Fardy Walsh, a tenant of Sir Henry Burke, were mutilated on two separate occasions. He had joined the Plan but had also paid his rent to avoid eviction. In February 1889, an unsuccessful attempt was made to blow up the house of Richard Martin, while, during the following August, the home of John Donoghue, a tenant on Lord Westmeath's estate, was razed to the ground. Martin and Donoghue had refused to join the Plan.[87] A Galway agent reported in April 1887 that on the property of William Ross Mahon the villagers were seeking the same reductions on the ground rents as the tenant farmers were on the agricultural ones. They were afraid of being boycotted. 'One sick man has paid the League his rent and has also paid William... He says he would lose more than the money in question if he did not pay them'.[88]

Boycotting was the most potent weapon in the nationalist armoury. Individuals who violated the code of the National League or the Plan of Campaign were mercilessly and rigorously boycotted. A Tipperary landowner attempted to define the process:

> It means that a peaceable subject of the Queen is denied food and drink and that he is ruined in his business; that his cattle are unsaleable at fairs; that the smith will not shoe his horse, nor the carpenter mend his cart; that old friends pass him by on the other side, making the sign of the cross; that his children are hooted at the village school; that he sits apart like an outcast in his usual place of public worship.[89]

Once the tenant farmers had resolved on the Plan of Campaign, the next step was to lodge the rents with trustees. The most notable and best publicised rent collection was on the Clanricarde estate in Galway. Lord Clanricarde had offered conditional abatements of twenty per cent to his non-judicial tenants but refused to reduce the judicial rents.[90] The tenants demanded an all-round reduction of forty per cent and the reinstatement of the four Woodford farmers evicted in August 1886. Their demands were refused and they adopted the Plan on 16 November. Two days later, as the agent Frank Joyce sat in the rent office at Portumna, Dillon and O'Brien arrived in the town and addressed a public meeting. A tenants' deputation called on Joyce and repeated their demands, which were again refused. 'They then said they were united and determined to carry out their programme and I answered that they were not a bit more determined in their plans than your Lordship and I were in ours'.[91] There then occurred what Dillon rather fancifully described as 'the most extraordinary scene that ever yet was witnessed in the course of the Irish land campaign'.[92] He and O'Brien were, musically and triumphantly, escorted to Keary's hotel and for the next five hours the tenants paid over their rents, less the agreed reduction of forty per cent.[93] Lord Clanricarde complained in the *Times* that his tenants were defrauding 'their long suffering creditor'.[94] Joyce demanded an immediate meeting with the Chief Secretary. 'If the government don't back us out in putting an end to this state of affairs and enforcing the payment of rents I don't know what can be done', he said.[95]

The Irish administration was slow to act and similar collections followed in many parts of the country. Lord De Freyne's Roscommon tenants paid over their rents to Willie Redmond on 2 December 1886.[96]

David Sheehy and Matt Harris organised the collections on the Galway estates of Lord Dunsandle and the Earl of Westmeath. John Dillon and Daniel Crilly were associated with those on Lord Dillon's huge Mayo property.[98] Crilly wrote from Ballaghdereen:

> The spirit of the people here is splendid. They are ready to go into any fight we ask them to enter. I never saw such evidence of pluck and determination anywhere as I saw on Wednesday at Charlestown and yesterday at Carracastle. Money came pouring in in both places.[99]

Publicans and priests, as well as politicians, did duty as rent collectors. The Tipperary tenants of T. J. Eyre paid their money to a local publican named Healy.[100] On the Clongorey estate in Kildare, about £130 was paid to Fr Kinsella and to a publican in Newbridge.[101] Fr Collins announced at the Kilbrin meeting in December 1886 that the Curass tenants had paid their rents to two priests who had lodged it with a trustee. 'I'm quite prepared to acknowledge that I received the money...The money is alright in Kanturk. Let them find it if they can'. Collins pledged that the money would be used to carry out the Plan of Campaign and 'to fight policemen and sub-inspectors, General Buller and all the rest'.[102] The parish priest of Dingle, Canon O'Sullivan, publicly announced that tenants on a neighbouring property had paid their rents to him, 'The money still remained in his hands', he said, and 'neither landlord nor agent had got it'.[103]

Rent collections, often daring, sometimes spectacular, had enormous publicity value. But more significantly, the payment of rent to trustees helped preserve the essential unity of the agitation. Divisional Commissioner Byrne noted:

> As the tenants who have joined the Plan have all paid money to some person whom they in many cases do not know and as they have got receipts which are legally worth nothing the fear of losing their money altogether binds them to adherence to the Plan no matter how strongly they may desire to shake it off and settle with their landlord.[104]

William O'Brien observed on more than one occasion that a tenant who abandoned the Plan was simply deserting his own money:

> 'From the moment the money is lodged in the estate fund', he said at Portumna on 18 November 1886, 'you need never bother your head any more about backsliders. You need not care a brass farthing from that moment whether this fellow or that fellow is

seen sneaking into the rent office after dark like a thief in the night. It won't matter a single pin because he won't have his rent to backslide with. His contribution to the common fund will be just as safe as if he were the bravest man on the property.[105]

A landlord had a number of remedies against a defaulting tenant. The most common was to proceed by writ or process as an ordinary creditor or to proceed by ejectment. In the former case, the sheriff was entitled to seize the tenant's stock or to sell the tenant's interest in the farm he occupied. To prevent seizures, farmers were advised to dispose of their cattle once writs had been served. The Limerick tenants of The O'Grady held a Plan of Campaign auction at Herbertstown on 14 December 1886 and the proceeds were lodged with trustees.[106] A campaign fair was held at Collon, County Louth, on 28 January 1887 and afterwards, according to the Irish Loyal and Patriotic Union, Lord Massereene and his agent were 'denounced and abused in unmeasured terms'.[107] The sheriff seized several head of cattle belonging to John Beytagh, a tenant on the Packenham-Mahon estate in Roscommon, and had them driven into Castlerea where they were advertised for sale. The cattle were purchased by John Fitzgibbon, the local National League leader, on the tenant's behalf. 'They were decorated with green ribbons', noted Sergeant Doyle, 'and driven in triumph through the town, escorted by a large crowd and several bands'. At the inevitable public meeting, a jubilant Fitzgibbon told the assembled farmers that, in their own peaceable and constitutional manner, they had 'squelched and battered landlordism' and prophesied that, within two years, the tenant farmers would be 'tilling and working the soil not in a misruled Ireland but in an Ireland emancipated and free'.[108]

Ejectment was the remedy favoured by a majority of landlords. A tenant who allowed himself to be evicted was promised financial support. 'Under the Plan of Campaign if men were put out of their lands they would be housed, clothed and fed as well as when they were in the enjoyment of their farms in prosperous times', guaranteed David Sheehy. 'This would continue until their broken landlords were plunged into bankruptcy by their creditors'.[109] The evicted 'would receive money to support them', promised Deasy, 'and they could go about the country visiting their friends while the landlords were whistling for their rents'.[110] The tenants were given to understand that eviction was simply a temporary inconvenience. 'As sure as the sun will rise tomorrow', pledged Dillon, 'they will get back into their homes'.[111]

A tenant who took a farm from which another had been evicted was vituperatively denounced as a landgrabber. David Sheehy clearly instructed his Galway audience on the treatment to be meted out to such an individual:

> You have only to avoid him in every way, to have no transaction with him. Tell everybody who he is. Describe him so that he may be known to every man, woman and child. Give a proper description of him to every man you meet that does not know him, the colour of his hair, length of his nose, whether he has long ears like a donkey or not. Tell them what kind of clothes he wears, what kind of coat he has. Let him be so described to every man, woman and child that no person can fail to recognise the creature as he walks through the street...I want you to put your foot down in such a way that he or his brother or any friends who shelter him—in fact to draw a line around these men—so that the people will stand away from him. If the people act in this way there is not a shopkeeper in Portumna, Tynagh, Woodford or Loughrea would sell him as much as a ha'porth of snuff and the ugliest woman in all Tipperary would not marry him.[112]

This system of exclusive dealing was not for the irrepressible member for Mid Cork. A punning Dr Tanner told his Tipperary listeners that there was 'nothing like leather and the toe of the boot strongly applied to the nethermost portion of the person in question is , as the English say, apt to make him "take his hook" at the shortest possible opportunity'.[113] 'Doctor' Francis Tully put his faith in stronger medicine:

> With reference to the grabber who holds this farm which belongs to the Widow Dempsey he used to have peelers around him and I was quite astonished to find that the police had left him. What has he for his protection now? He has two dogs. Well I hope and trust that I will be in a position to say from this platform when it will be erected again to reinstate the widow and orphans... that these two dogs bit him on the roadside.[114]

(v)

In January 1888, the police estimated that the Plan of Campaign was in operation on 38 estates.[115] In the following month, Dillon informed the members of the House of Commons that the agitation had been successful on 42 of the 95 properties on which it had been introduced.[116] When he gave evidence before the Evicted Tenants Commission in

January 1893, he cited 116 instances of the Plan of Campaign.[117] During its five-year history, the Plan was, in fact, carried out on at least 203 properties, 75 in Munster, 71 in Connaught, 33 in Leinster and 24 in Ulster, 10 of the latter being in Donegal alone. Munster and Connaught accounted for more than 70 per cent of all Plan of Campaign estates. The heaviest concentration was in the congested districts along the Atlantic seaboard.[118]

The problems confronting Irish farmers in the mid-1880s were magnified in the more disadvantaged western counties, where poor land and small holdings increased the vulnerability of tenants to economic and climatic changes. They were affected earlier and more consistently than individuals who lived in the eastern half of the country.

The property of Herbert Baldwin O'Sullivan, consisting of 2,400 acres divided into thirty eight holdings, was situated in Muskerry West, about seven miles from Macroom in County Cork. An inspector of the Land Commission reported in March 1890 that this land was all of a mountainous character with bleak and almost barren soil.

> Ridges of rock, almost sterile, and large tracts of mountain moor and peat bog are the chief characteristics of the estate. The stock on such lands, from the very nature of things, are small and the first to feel a drop in price and the last to rise in value. Bran and meal are given to cattle even in summer to supplement the unnutritious grass.

The land was suitable only for pasture and the fall in butter prices in recent years had greatly affected the tenants.[119]

The Browne property at Timicat, County Galway, consisted of 486 statute acres, of which 176 were 'unallotted waste'. The farms of the 19 tenants ranged from 4 to 34 acres. A constabulary inspector reported in April 1888 that the land was cold and wet and largely reclaimed bog.

> 'The tenants on this property bear the character of being an industrious, hardworking and peaceable people,' he wrote. 'Every year the adults go to England and it is the money thus earned that pays the rent. Their holdings of bad, wet land would never support them'.[120]

Similar reports on the Nixon, Stewart and Olphert estates in Donegal and the Dillon and Taffe properties in Mayo suggest that the situation confronting the Browne tenants was the norm rather than the exception in the congested districts.[121]

The commission of inquiry into the Irish land system which the Salisbury government established in mid-1886 defined farmers as those wholly dependant for their livelihood on the produce of their holdings. Under this definition, the inhabitants of the congested districts could not be classified as farmers. On the contrary, the report stressed, they must be regarded 'as labourers who occupy residences with portions of land attached which assist them in the support of themselves and of their families'. Rent was paid, not out of the produce of the land, but from wages earned as seasonal labourers in England and Scotland or by fishing or by working for farmers at home. The acute economic depression of the mid-1880s reduced the prospect of finding work locally, while the introduction of mowing and reaping machines into England and Scotland lessened the demand for seasonal migratory labourers from the west of Ireland.[122] The 80,000-acre estate of Lord Dillon in County Mayo is a case in point. The property was let at twenty five per cent below the government valuation of £20,000. Each of the 5,000 tenants paid rent of less than twenty pence an acre.[123] Yet a visitor to the estate in the summer of 1887 claimed that 'if the people were living rent-free the land would not support them for six months of the year'. They were used to earning good wages in England, he said , but at present their earnings were fifteen to twenty per cent lower than usual.[124]

Should purchase become an option for occupiers in the congested districts, the Land Commission was chary about advancing loans on the security of their holdings. The Taffe estate consisted of 9,000 acres in Mayo and 1,300 in Galway. The rental of £4,400 was reduced to £3,543 after the passing of the Land Act of 1881. The estate was mortgaged to the extent of £39,000 and was placed in Chancery in 1880. After the passage of the Ashbourne Land Act in 1885, the tenants were given the option of acquiring their holdings at fifteen years purchase of the judicial rents. They accepted and requested the Land Commission to purchase the estate on their behalf. The Commissioners refused and, even when the purchase price was reduced to thirteen years, they claimed the transaction could not be completed without loss to the tenants. 'The bulk of the holdings were occupied by a large and poor population', the Commission noted, 'whose means of support are not derived from their holdings or local employment but from wages earned in England or elsewhere'.[125]

The economic depression of the mid-1880s affected the price of all types of produce in all parts of the country but its impact was less severely felt outside the congested districts. In the midlands, east and south-east, tenants were less homogeneous and the multiplicity of reasons put forward for adopting the Plan of Campaign suggests a diversity of interests and aspirations.

Peter de Penthony O'Kelly's property at Clongorey, County Kildare, was subject to periodic flooding. In October 1886, the river Liffey burst its banks. Many houses in Clongorey were destroyed and crops were swept away. It was this disaster that prompted the tenants to seek rent reductions and, when these were refused, to adopt the Plan.[126] The tenants of John Christopher Delmege, in Glensharrold, County Limerick, were described by County Inspector Roberts as miserably poor and wretchedly housed, while the land was of the worst description.[127] A special committee, appointed by the Limerick Board of Guardians, found that Glensharrold was mainly 'wild and barren, mountain and bog in its prehistoric state'. The general condition of the people, the report concluded, was one of chronic poverty and misery.[128]

The tenants of Mr Bannon in King's County adopted the Plan because they objected to the newly appointed agent, a Mr Garvey, who had 'the name of being "sharp". The tenants had a wish for the former agent', noted Divisional Commissioner Butler.[129] The dispute was settled in December 1889 and the tone of a leading article, entitled 'The Broughal Slaves', in the local *Midland Tribune*, suggests that the tenants had been less than committed to the ideals of the Plan of Campaign:

> They have forfeited all claim to sympathy in the dark hour which, we can assure them, is not far off. Their doom is sealed... They have branded themselves as rogues and the hand of every honest man will be raised against them for all time to come.[130]

Agents appear to have provoked or exacerbated a number of disputes. It was suggested that the Plan was started on the Brooke estate in County Wexford because the energetic director of the Property Defence Association, Captain E. C. Hamilton, held the agency.[131] In December 1886, Lord Massereene appointed Athol Johnston Dudgeon, a partner in a Dublin firm of solicitors, as agent.

> 'There has been no peace on the estate since then', reported H. Thynne, Deputy Inspector General of the Royal Irish Constabulary, in April 1889. 'It is the general opinion that things will never settle down on the Massereene property so long as the management is left to an absentee firm of solicitors whose first care is to accumulate costs'.[132]

In November 1887, E. Whitby Lynch resigned as agent and receiver of the Taffe estate in Mayo. His successor, D. J. Hughes of Kiltimagh, was bitterly opposed by the tenants and they determined not to pay any rent as long as Hughes held the agency.[133] A speech delivered by William O'Brien at Killeagh, County Cork, in December 1886, suggests that there was widespread dissatisfaction with land agents:

> One of the beauties of the Plan of Campaign is that before many months are passed it will abolish the hateful trail of a foul class of men who have the vices of aristocrats with the blood of plebians and the hearts of vultures, the class of village tyrants who for many a day and many an age have lived and fattened upon the suffering and misery and life blood of the people.[134]

No doubt harsh and exacting landlords, like the infamous Lord Clanricarde and the equally notorious Colonel O'Callaghan of Bodyke, County Clare, existed in every Irish county and, if such landlords were loath to reduce their rents, their tenants were not hesitant in following Willie Redmond's advice to give them 'a clout of the Plan of Campaign'.[135] While leading advocates of the agitation repeatedly stressed that it was only to be used against unjust landlords, the evidence suggests that the majority of estates on which the Plan was implemented were carefully chosen and that landlord vulnerability, rather than obduracy, was the determining factor. Landlords foundered between Scylla and Charybdis. Few could afford to grant rent reductions ranging from twenty to fifty per cent but, if they refused, they were unable to pay family charges and they became subject to penal interest rates on their mortgages. If they bankrupted themselves in the process, as a visitor to Lord Dillon's estate in June 1887 prophesied,[136] the tenants alone would be the beneficiaries.

Daniel O'Connell of Tulla, County Clare, Mrs Gibbons of Kilmovee, County Mayo and Mr Browne of Timicat, County Galway, were small property owners who shared a common dilemma. O'Connell was 'miserly poor', Mrs Gibbons 'was pressed for means', while Browne complained to the Chief Secretary in March 1888 that his income and resources were limited. He could not afford, he wrote, to evict his tenants, garrison the townland with emergencymen and do without rent, while being subject to taxes and other charges, for an indefinite period, unless the government came to his aid.[137] The result was invariably the same. Hard pressed owners held out for a time against the Plan and then surrendered, frequently ceding the tenants' full demands and generally earning the detestation of the police. 'Landlords have shown great weakness and folly in many cases', Divisional Commisioner Byrne charged, 'by first refusing the tenants' demand but eventually yielding after costs and time had been expended'.[138]

Debts, whether inherited or incurred, family charges, such as jointures or portions for sisters and younger brothers, and yearly interest payments on crushing mortgages, drained even the most handsome of incomes. Hastings Twiss succeeded his brother at Bird Hill, County Tipperary, in the late 1870s. He and his sisters were annuitants, each in receipt of £2,000. The sisters had a prior claim and, to secure their annuity, they placed the property under the control of the Court of Chancery. Over the previous eight years, his sisters had received their full share but Twiss had been paid less than one third of the £16,000 due to him. He lived on the estate but the property was managed by trustees and he was powerless to interfere with its running. His sisters lived abroad.[139]

In the late 1860s and early 1870s, land was regarded as a very sound security and financial and religious institutions alike invested heavily in Irish property. After the disestablishment of the Church of Ireland in 1869 the Representative Church Body loaned almost half of its £8.5 million compensation to Irish landlords.[140] Colonel C. G. Tottenham received £16,000 in 1873 and Lord Dunsandle got £100,000 in 1871. The multi-troubled Viscount Massereene and Ferrard, who was described by a high-ranking police official as 'a man of intemperate habits' and, less coyly, by a visitor to his estate as 'a poverty-stricken man and a somewhat drunken character', was advanced £120,000 between January 1872 and December 1878. When the Countess of Kingston decided to consolidate the various charges and mortgages on her estates, she was loaned a staggering £236,000 in 1875.[141]

Maynooth College, which also received compensation under the 1869 Disestablishment Act, granted mortgage loans to seven Catholic landowners. A sum of £44,870 was advanced to Baron De Freyne, of County Roscommon, in May 1878, while an additional £13,630 was sanctioned between January 1879 and June 1885.[142] Between April 1872 and June 1876, Lord Granard, a convert to Catholicism, mortgaged lands in Leitrim and Longford, including the castle and domain of Castleforbes, to the trustees of Maynooth in order to secure almost £94,000. There was a clause in the contract which prevented the college from foreclosing on condition that the interest was paid within two months of the gale day. By 1887, Lord Granard was hopelessly in arrears and, in March, he wrote to Dr Browne, the president of Maynooth, to apologise. 'It is of course from no fault of mine that this has occurred', he protested, 'and I can most thoroughly sympathise with you as I do not receive a shilling from the estate at present'. By the autumn, the arrears of interest had swelled to some £5,470.

On 18 October 1887, the trustees decided to dispose of the estate but agreed to defer their decision so as to give Granard an opportunity to discharge his debt. A fortnight later, he lodged a sum of £1,000 to the credit of the college and the trustees gave him until Febuary 1888 to clear the outstanding arrears. On 30 January, Granard's solicitors informed Dr Browne that they had failed to raise a public loan. 'Owing to the present state of affairs', they wrote, 'it is almost impossible to borrow money on any landed security in Ireland'.[143] However, Lord Granard's relations agreed 'to help him in the present emergency'. The trustees were asked to forgo £1,000 and to accept £3,500 as a full settlement of the outstanding arrears. They refused and, almost immediately afterwards, the estate was placed in Chancery, a receiver was appointed and the property was offered for sale to the tenants.

In June 1888, P. A. Chance, solicitor to the Maynooth trustees, discovered that Granard had received a net rental of £19,000 from his estates for the two years up to May 1885. Between that date and December 1887, some £15,000 was collected, giving the lie to Granard's repeated assertions that he had received no rent since the land agitation commenced. In these four and a half years, according to Chance, the tenants had paid 'far more than the yearly valuation' which led him to conjecture that the rents were applied to pay the interest to puisne mortgagees.[144]

Justice Monroe asserted that the Langford tenants of Meelin, County Cork, had adopted the Plan of Campaign because the estate was a poor one and the landlord was known to be practically insolvent.[145] Many Plan estates were in notoriously straitened circumstances. Several were under the control of the Court of Chancery, including the Cork properties of H. B. O'Sullivan and Sir Joseph N. McKenna, MP, the Taffe estate in Mayo and that of J. C. Delmege in County Limerick. The latter landlord purchased the Glensharrold property in the Encumbered Estates Court for £8,000 and subsequently mortgaged it to the Scottish Union and National Insurance Company. The estate was placed in Chancery in 1884 after Delmege had fallen into arrears with interest payments. In April 1888, he was sued in the Court of Exchequer for £9,877 by the mortgagees and a decree for the full amount and costs was given against him.[146] The O'Kelly property at Clongorey, Co. Kildare, was heavily mortgaged and was managed by trustees,[147] there were 'very large charges' on Lord Dillon's Mayo estate,[148] while the Earl of Kenmare's property was said to be irretrievably 'in the red'.[149] Lord Milner visited Kerry in the autumn of 1886 and was introduced to the Earl's prospective daughter-in-law, an heiress named Miss Baring. He noted tartly that she had a decided role to play, not alone in replenishing the family coffers, but in strengthening the family character.[150]

The adoption of the Plan of Campaign by the Lansdowne and O'Grady tenants should dispel any remaining doubts as to the essential nature and purpose of the agitation. The conflict on these two properties strongly suggests that, in the more prosperous areas of the country at least, the main determinant in adopting the Plan was tenant opportunism based on the landlord's known financial vulnerability.

The police maintained that the agitation on The O'Grady's property at Herbertstown in County Limerick was probably the least justifiable of all those started in 1886-7.[151] Wilfrid Blunt was informed by a Limerick priest that the struggle had done great damage to the nationalist cause. Fr Higgins stated that the tenants were not rack-rented and that The O'Grady was the best landlord in the county.[152] In August 1887, The O'Grady issued a circular in which he stated that he was a resident landlord whose ancestors had lived at Kilballyowen for more than 400 years. He knew every tenant personally and friendly relations had always existed between them. Evictions were unknown on the estate. In 1885, recognising the fall in price of stock and produce, he reduced the judicial rents by fifteen per cent and the non-judicial ones by twenty five per cent. In October 1886, similar abatements were offered but the tenants demanded an all-round reduction of forty per cent. This he rejected, as he considered it to be 'unreasonable and unjust', whereupon the tenants adopted the Plan of Campaign.[153] The estate was very heavily encumbered. The annual interest on mortgages amounted to £1,200; drainage charges came to £110, head and tithe rents to £174; the dowager Madam O'Grady held a jointure of £500. The gross rental of the estate was £1,742, which produced a net income of £1,304, according to the agent, after abatements, bad debts, poor rates, county cess, etc. were deducted. The income which The O'Grady derived from farming his 400 acre demesne was essential to make up the annual shortfall.[154]

George Pellew, an American lawyer who spent four months in Ireland during the summer of 1887, visted Herbertstown in August and was informed by the secretary of the local National League branch, a man named McGuire, that the property was heavily mortgaged:

> O'Grady handled only five or six per cent of all the rental of Herbertstown. The encumbrancers will have to foreclose and then put the estate up for sale and there will be no one to buy it but the tenants.
> 'The result will be', I hinted, 'that you will buy the land yourselves pretty cheaply'.
> 'Yes, that is what we all expect', McGuire replied.[155]

In October 1886, Lord Lansdowne reduced the judicial rents on his Kerry estate by twenty per cent and offered reductions ranging from twenty five to thirty five per cent to his non-judicial tenants. The representatives of his Queen's County tenants demanded similar abatements. Lansdowne refused to reduce these judicial rents as he felt that the circumstances on his Kerry and Queen's County estates were fundamentally different and that farmers on the latter were less affected by the fall in prices. They consequently adopted the Plan on 17 November 1886.[156]

Lord Lansdowne, who was appointed Governor-General of Canada in 1883, owned a palace in London and estates in England and Scotland, as well as 95,000 acres in Kerry, 13,000 in Meath and 9,000 in Queen's County. But he was, as his biographer wryly pointed out, in serious financial difficulties. His father had left the Scottish property to his widow for her lifetime and debts of £300,000 to his son. Family charges and mortgages were a crippling annual burden and the estate in Wiltshire provided little income.[157]

Lansdowne complained continuously of the financial drain of the Queen's County dispute. 'The expense is enormous', he wrote to his mother in July 1887, 'horses, cattle, labourers, agricultural machinery, provisions brought from a distance'. In December, he reported haplessly: 'If my Luggacurran tenants do not come to terms I shall have to continue a state of siege there for an indefinite time... I really don't know where the sinews of war are to come from and there will have to be another raid upon the pictures'. His financial exigencies induced him to accept the Viceroyalty of India, which the Prime Minister had offered him on New Year's Eve, 1887. India would help reduce that 'load of debt which has been so terrible an incubus to us all' and provide an alternative to 'living in a corner of the house in England, perpetually worried by financial trouble and perhaps increasing, instead of diminishing, the family liabilities'.

Lansdowne's position vis-à-vis the Plan of Campaign was exceptional. As an office holder under the Salisbury Government, he could hardly surrender to an agitation which the courts had pronounced illegal and the Tories had denounced on moral grounds. Despite the cost, he had no option but to continue the struggle. By mid-October 1887, after the Plan had been in operation for almost a year, Lansdowne concluded that the only solution to his problems, and to those of Irish landlords generally, lay in transferring the ownership of the land to the occupiers. 'It would be an immense relief', he wrote, but he hoped that the terms would stop 'short of confiscation'.[158]

Denis Kilbride and John Dunne were two of the signatories to the appeal for rent reductions in November 1886. The former held 700 acres and paid an annual rent of £760, while the latter farmed more than 1,300 acres. Shortly after they were evicted in March 1887, Kilbride became a nationalist MP for a Kerry constituency and Dunne entered one of his horses for the Curragh races.[159] At a meeting of the central branch of the National League in Dublin on 29 March 1887, Kilbride clearly stated the position of the Lansdowne tenants: 'The Luggacurran evictions differed from most of the other evictions to this extent that they were able to pay the rent... It was a fight of intelligence... against intelligence... It was diamond cut diamond'.[160]

A tenant holding four acres of Connemara bog had fundamentally different problems and aspirations from one who kept racehorses and farmed 1,300 acres in the fertile Queen's County. Along the Atlantic seaboard, the Plan of Campaign was a struggle for survival by impoverished peasants. In the midlands, east and south-east, tenant farmers also felt the pinch of depression but they took advantage of it to exploit financially vulnerable landlords. Virtually all of the crucial Plan of Campaign struggles were fought on estates which had notoriously insecure financial bases.[161] If the tenants, by withholding rents, could bring landlords to the verge of bankruptcy and force them to sell at sacrificial prices, the tenants would be the beneficiaries. From the perspective of middle-class politicians, the Plan of Campaign was essentially a political conflict rooted in economic difficulties; for the peasants of the congested districts, it was a matter of survival; the tenant farmers from the more prosperous areas of the country saw it essentially as a class struggle directed against property owners. The Plan of Campaign reflected the hardships and aspirations, whether political, social or economic, of a far from homogeneous people.

In mid-January 1887, the Ulster Unionists' parliamentary leader informed the Orange card-sharp that

> very bitter feelings are rising up amongst the northern loyalists owing to the fact that the lawless portion of the community are allowed to have everything their own way. If it should continue to be shown that disloyalty and lawlessness mean, amongst other things, immunity from paying just debts, I need hardly point out how popular disloyalty and lawlessness are likely to become.[162]

Colonel Saunderson, or 'the dancing dervish' as he was dubbed by the Liberals,[163] was, however, being unduly alarmist. The parliamentary commission of inquiry into the Irish land system, which reported early in 1887, concluded that combinations against the payment of rent and for the right of free sale did not as a rule exist in Ulster.[164] One of those interviewed, Charles Pringle, a tenant farmer from Caledon, County Tyrone, stated:

> There is no combination of any kind against the payment of rent that I am aware of in the north. The tenants do not meet together and come to an agreement, saying, 'we propose that we will pay a certain amount and if that amount is not accepted we will pay none'. There is nothing of that sort here that I know of.[165]

Seventeen of the twenty four Ulster properties on which the Plan of Campaign was adopted were in Donegal, Cavan and Monaghan. Derry and Fermanagh accounted for one each and there were two instances in Armagh. On one of the three Tyrone properties involved, more than half the tenants refused to join and paid their rents in full.[166] In December 1886, it was reported that Antrim farmers were 'too wary and clever' to become involved, while the tenants from County Down, although affected by the fall in prices, were paying their rents.[167]

The northern farmer's concept of himself as the 'honest Ulsterman' was deep rooted.[168] Walter Osborne, a judicial tenant holding 108 acres in Campsey, County Derry, informed the Cowper Commission: 'We in the north do like to pay our debts'.[169] According to Samuel McDowell, from Carrickfergus, Ulster tenants were 'a very law abiding people...not easily taken up with by agitators'.[170] The Irish Society's agent, Thomas Montgomery, a southerner, stated that northern tenants were 'far more inclined to pay and far more inclined to discharge their liabilities' than their southern counterparts.[171] Dissenting Protestants, in particular, felt an obligation to fulfil their contracts. John Lane, a holder in perpetuity under the Duke of Abercorn, claimed, in late October 1886, that Ulster rents were being well paid and that boycotting was not a feature of the province. 'I do not think that amongst a Presbyterian population you will ever see anything of that sort', he opined. Lane categorised Presbyterians as 'beasts of burthen' who invariably attempted to discharge their obligations.[172]

In the wake of the seventeenth century plantations, a special relationship developed between the mainly Anglican landed gentry in Ulster and their Protestant tenants, be they Presbyterians or members of the Church of Ireland. A major bonding factor had been the general recognition of tenant right by the landlords of the province. Tenant right, or the Ulster custom as it came to be called, defied definition. There were as many variations and interpretations of the custom as there were landlords and tenants. Its essence, however, was the acknowledged right of the tenant to sell his interest in his holding, which was defined as the value of the occupancy right.[173] The Ulster custom, northern commercial and industrial prosperity generally, and the linen industry in particular, made Ulster farmers, in the words of Lord George Hamilton, son of the Duke of Abercorn, 'the best conditioned people in the United Kingdom'.[174]

Northern farmers deeply cherished their customary rights of tenure. John Hancock, Lord Lurgan's land agent, informed the Devon Commission in 1845 that the Ulster custom was 'one of the sacred rights of the country which cannot be touched with impunity, and if systematic efforts were made amongst the proprietors of Ulster to invade tenant right, I do not believe there is a force at the disposal of the Horse Guards to keep the peace of the province'.[175] It was, however, the Ulster custom's lack of homogeneity and the wide variations in its application that led to friction between northern landlords and their tenants, particularly during the 1870s and 1880s. Gladstone's Land Act of 1870, by giving the force of law to the Ulster custom as it existed, institutionalised practices with which many tenants were unhappy. An additional cause of tenant discontent was the unwillingness of northern landlords to dispose of their estates under the purchase provisions of the 1881 and 1885 Land Acts. The Ulster custom had advantages for landowners as well as occupiers. By enhancing tenant security, it secured the landlords' rents and, consequently, the owners were reluctant to sell. Andrew Kennedy asserted before the Cowper Commission that 'tenant right is the security for the rent.'[176] John Lane, questioned as to the willingness of northern landlords to sell at a fair price, replied, 'Oh, no! the rents are too well paid for that'.[177]

The constitutional crisis of the mid-1880s swamped the Ulster tenants' economic grievances. Gladstone's abortive attempt to introduce Home Rule restored the pre-land legislation status quo between landlord and tenant. The reaction of Ulster Protestants — be they Anglican or Dissenter, landowner or occupier, industrialist or worker — to Gladstone's bill was almost universally hostile.

The Plan of Campaign, which was an exclusively Catholic movement, was, according to its opponents, politically motivated. Far from being a rent agitation, the *Times* editorialised, the Plan of Campaign was 'a deliberate attempt to break down law and sweep away property in order to accomplish the overthrow of the imperial authority in Ireland'.[178] The agitation's leading opponent, Arthur Hugh Smith-Barry, informed his English constituents that 'the Plan of Campaign had really nothing to do with landlords and tenants but was a great political movement started by the leaders of Irish rebellion to prevent the possibility of government being carried out in Ireland by the English government'.[179] The nationalist political leadership and the Irish Catholic Church, at both hierarchical and parochial level, made little or no attempt to allay Protestant fears. The leading proponents of the Plan of Campaign made no secret of the fact that their ultimate aim was the establishment of a Home Rule parliament in College Green.[180]

A mass meeting of Lord O'Neill's County Antrim tenantry was summoned for Randalstown on 15 December 1886, after which the tenants' executive committee issued the following warning:

> A course is being pursued by a few misguided men in the district calculated to bring upon County Antrim and our peaceable neighbourhood the curse of agrarian agitation. They have adopted the most vicious principle that lies at the root of the Plan of Campaign, namely, that the tenant is justified in fixing the rent he is to pay and if the decision arrived at is not bowed to by the landlord that agitation follows to coerce acquiescence... It has been openly disclosed by the leaders in the South that their object is to render the government of Ireland impossible; and the action proposed in Randalstown is calculated to strengthen their hands.

The committee urged 'every honest and loyal man' to attend and to give moral support to the government in its efforts to restore peace and order to the country. 'By your numbers, your presence and your voice let it be known that the Plan of Campaign will not flourish on the soil of County Antrim'. Orange coloured notices, urging that the Randalstown demonstration be turned into a 'loyal meeting of honest men' were extensively posted. 'Let it be known', they proclaimed, 'that you too have a Plan of Campaign! Orangemen and honest men to the front — the eyes of Ulster are upon you'.[181]

On 14 December 1887, the *Freeman's Journal* carried a report under the heading, 'Orange tenants and the Plan of Campaign: an Antrim landlord caves in'. The tenants on the Joy estate near Ballymena threatened to adopt the Plan after their demand for a fifteen per cent rent reduction had been refused. The landlord immediately surrendered and granted the abatement. The account concluded: 'Mr Joy is the representative of a good old family and the tenants are almost exclusively Orangemen'.

Joy, who was not the owner but the agent, refuted the newspaper report. He wrote to his friend, District Inspector Crohan of Ballymena:

> I should be sorry to think that the Plan of Campaign had any influence over any tenants I have to deal with and I am thankful to say that such an illegal and immoral plan has no acceptance with them and so far as I know they never alluded to it in any but a joking way.

Crohan subsequently informed the Northern Divisional Commissioner that 'the Plan of Campaign had never been resorted to in this locality'.[182]

Home Rule and the Plan of Campaign had at least one northern Protestant supporter. The Rev Matthew Macauley, the seventy-year old Presbyterian minister of Castleblayney, County Monaghan was an advocate of tenant right and a Gladstonian Liberal who publicly endorsed the Home Rule ticket. For his trouble, he and his wife were assaulted on the public highway by an Orange drumming party on 1 July 1886. A week later his church was desecrated; the words 'no Home Rule' were suspended from the pulpit and orange lilies were scattered about.[183] On 30 November 1886, a monster meeting was held at Castleblayney's Fair Green to consider the Plan of Campaign. Canon Hoey presided and the speakers included John Dillon and Willie Redmond. Macauley, in a letter apologising for his inability to attend, wrote: 'I hope you get good weather and that you will hear rousing addresses from the patriotic gentlemen whose names are on the placards. I hope also that there will be a productive meeting and that the Plan of Campaign will be adopted'.[184] Macauley was subsequently forced to resign his ministry and to leave the area.[185]

4 *'Mickey-the-Botch'*

The difference of moonlight and moonshine
The people at last understand;
For moonlight's the law of the League
And moonshine's the law of the land.
Samuel Murray Hussey, *Murray's Magazine,* Vol. 1, May 1887

It is a terrible state of affairs to live in and all due to weak government.
Arthur McMurrough Kavanagh to Madam Elizabeth O'Grady, 16 February 1887. (NLI Ms.22461, The O'Grady Papers)

After the Tories had been overwhelmingly returned to power, following the defeat of Gladstone's Home Rule Bill in the summer of 1886, a less than enthusiastic Sir Michael Hicks Beach was appointed Irish Chief Secretary. He differed fundamentally from the Prime Minister and his cabinet colleagues in his attitude to Ireland. In sharp contrast to Salisbury's often repeated declarations in favour of restoring law and order and of enforcing property rights, Hicks Beach saw his task as the 'humbler' one of 'keeping things going'.[1] He recognised that wholesale evictions were a grave threat to the peace and stability of the countryside and he was opposed to taking strong executive action on behalf of Irish landlords whom he disliked and despised.[2]

From the beginning of September 1886, the Chief Secretary was greatly influenced by General Sir Redvers Buller, lately dispatched to the south-west of Ireland as a 'special commissioner' to suppress boycotting and outrage. Within a week of his arrival, Buller reported to Dublin Castle that tenants in Kerry and Clare were impoverished and were being evicted at will by rack-renting and impecunious absentee landlords. As a result, crime was endemic and the people had lost all respect for the law. Buller concluded that the land system was the root cause of all the problems and the solution could only be found by the government settling the land question in a way that was satisfactory to the 'majority of the occupiers'.[3]

The constant stream of reports he received from the south-west convinced the Chief Secretary that the executive would have to intervene directly to curb evictions. He proposed that government officials should approach landlords privately and urge them to act responsibly and with greater restraint than they had shown heretofore. If the landlords refused, Hicks Beach was prepared, as a last resort, to refuse military and police assistance to the sheriff in carrying out evictions, although he stressed on 18 September that circumstances were not as yet sufficiently grave to warrant such an extreme step. In the meantime, he suggested that government lawyers should plead the tenants' case at eviction proceedings. After this proposal had been rejected out of hand by the Attorney-General, Hicks Beach made the extraordinary suggestion that, in a notoriously bad case, Buller should use the funds at his disposal and secretly engage a good solicitor to act for the tenant. 'The case must be a really bad one to justify such interference', warned Hicks Beach, 'though you might use the possibility of it as a threat to make the landlord do what you thought right, without much risk'.[4]

The success or otherwise of the Chief Secretary's policy depended on a number of imponderables — the character, temperament and generosity of individual landlords, not to mention their financial position and political commitments. Hicks Beach's difficulties were neatly pointed up by the contrasting attitudes of the Marquis of Lansdowne and the Marquis of Clanricarde.

Lansdowne was the financially-pressed Governor-General of Canada and, as an office-holder under the Crown, he felt obliged to publicly associate himself with the government's Irish policy. In October 1886, he granted considerable abatements to his Kerry tenants and among the reasons he cited were the falling agricultural prices, the poor quality of Kerry butter and livestock, the collapse of the Munster Bank and the general restriction of credit. He emphasised that he was responding to 'the strong appeal which had been made to the landlords by the leaders of political opinion on both sides . . . to diminish the difficulties of the coming winter'. He believed that the simplest way to do this was to reduce the rents.[5]

The wealthy and eccentric Lord Clanricarde proved less amenable to government pressure. The whole of south-east Galway had remained seriously disturbed in the aftermath of the evictions that had been carried out on his estate in late August 1886 and the prospect of a renewed land agitation was ever present. The Chief Secretary interviewed Francis Joyce, Clanricarde's agent, in early October and warned that, unless Clanricarde reduced his rents, the government would do all in its power to obstruct further evictions on the estate.[6] This meeting produced a petulant letter from Clanricarde which

SUPPLEMENT GIVEN AWAY WITH THE **WEEKLY FREEMAN** 17TH DECEMBER, 1887.

"THE MOST NOBLE."

Counsel for The Most Noble the Marquis of Clanrackrent.—"Gentlemen of the Jury, I put it to you, is my client as black as he is painted?"

Jury unanimously.—"Yes! Blacker!!!

gave Hicks Beach the opportunity to make a comprehensive and explicit reply. The Woodford evictions, he wrote, were expensive and troublesome but the government had indicated its determination to enforce the law. In return it was expected that landlords would act reasonably and responsibly. Clanricarde had been advised by his agent to grant considerable abatements but had consistently refused to do so. Hicks Beach warned that, if Clanricarde continued to insist on the letter of the law, he would find such a course both costly and ineffective. A request for police or military aid at evictions 'would certainly be retarded by the pressure of other claims and duties and would most probably be postponed to the utmost extent permitted by law'. The Chief Secretary, in conclusion, advised Clanricarde to follow his agent's advice and to display some leniency towards his tenants. Clanricarde, however, was uncompromising and the agitation on his estate was not finally settled until 1913.[7]

The intractability of landlords like Clanricarde would ultimately and inevitably render Hicks Beach's policy ineffective but, up to the end of October 1886, he believed that his attempts 'to make landlords reasonable' were succeeding. On 20 October, he wrote to the Prime Minister:

> Outrages have been curiously decreasing in the past three weeks... From all parts of the country, I hear reports of a better tone among the people. Landlords are giving considerable reductions of rent and tenants seem less unwilling to pay when they can. Some of the Parnellites are stumping the country, but their speeches are more moderate in tone than formerly and in some of the worse parts... their agitation has fallen flat.[8]

The pressure being applied to the landlords by the Irish administration could not be concealed from the public indefinitely and on the very day that the Plan of Campaign was published *United Ireland* leaked a correspondence, dated 14 October 1886, between Sub-Sheriff Gale of Cork and County Inspector Moriarty of Kerry, in which the latter refused to provide police protection at evictions unless ten days' notice was given. Moriarty also demanded that full details of the evictions be furnished to enable police inquiries to be made.[9] It was reported that General Buller himself was investigating each case and demanding the abandonment of evictions if he found that tenants were unable to pay.[10] The *Freeman's Journal* claimed that the tenants were in receipt of indirect but unequivocal government support and were being treated leniently and sympathetically by the constabulary and judiciary alike.[11]

The English liberal press accused the government of vacillation and confusion, and charged that the policy which Hicks Beach was enforcing in Dublin contradicted the one Salisbury was expounding in London. The Irish executive was taxed with using a 'dispensing power', i.e. of refusing to implement the decisions of the courts. According to the liberal press, such actions justified the policies of the National League and, in particular, Parnell's Tenants Relief Bill. The government has 'strained at the gnat and swallowed the camel', opined the *Birmingham Daily Post.*[12]

The nationalists claimed that they and the Irish executive, 'though travelling by different trains, were destined for the same station. Buller was hailed as the saviour of the downtrodden peasantry. The ebullient MP for Mid Cork, Dr Charles Tanner, compared him to the atheist who went to church to laugh at the sermon: 'He came to scoff but he remained to pray'.[13] Dillon suggested that there was in existence 'a ministerial Plan of Campaign' and confessed himself surprised 'to find Sir Michael Hicks Beach and the Tory executive as our allies'.[14] The inimitable William O'Brien informed Lord Clanricarde's tenants that the actions of the Irish executive justified the Plan of Campaign:

> The government themselves are exerting all sorts of pressure to extract from the landlords down in Kerry larger reductions than you are looking for from Clanricarde. You are only assisting the government by trying a little gentle pressure of your own. You have no General Buller here in Galway. You have only to try and do a little Bullerising on your own account.[15]

The government was greatly embarrassed. Buller, Salisbury and Hicks Beach successively, and unconvincingly, denied the use of a 'dispensing power' in Ireland. At the Lord Mayor's banquet in London on 8 November 1886, the Prime Minister equivocated: 'We possess no dispensing power; we have exercised no dispensing power; we have merely tried to administer the law as we found it'. He finally admitted, however, that the government had privately exhorted Irish landlords to act reasonably towards their tenants.[16] The nationalist press relished the nuances:

> Men learned in distinguishing between Tweedledum and Tweedledee may find a vast difference between the exercise of the 'dispensing power' and the exercise of 'exhortation' but the tenants are quite as well satisfied with the latter as they could be with the former.[17]

The obduracy of many landlords and their inability or unwillingness to yield to government pressure fostered the Plan of Campaign. The O'Grady tenants in County Limerick were the first to adopt the Plan, a fortnight after its promulgation in *United Ireland* on 23 October, and their example was eagerly followed by farmers throughout the country. The Chief Secretary recognised the potential danger of the new agitation and, before his policies became completely ineffective, he and Buller made a last-ditch effort to recover the situation and to reconcile landlords and tenants before it was too late.

On 1 November 1886, Buller, whose temporary appointment as a special commissioner in the south-west was drawing to a close, made an extraordinarily outspoken submission to the commission of inquiry into the Irish land system, which the government had established some months previously. Buller asserted that the law was biased in the landlords' favour and that, in the south-west, even the judicial rents were too high:

> You have got a very ignorant poor people and the law should look after them, instead of which it has only looked after the rich... There is not much law in this part of the country but what law there was was really on the side of the rich.

He maintained that the inequitable legal system and high rents had alienated the tenants and forced them to resort to the National League, which they regarded as their only salvation.[18]

Two days later, addressing his constituents at Bristol, a less emotional Hicks Beach echoed Buller's sentiments. He maintained that great progress would be made if the Irish people could be convinced that 'the reign of law meant the reign of justice to rich and poor alike, that the rights of the poor man as well as the rich would be respected and that the law should never be used as an instrument of oppression'. The Chief Secretary denied that the Irish executive was exercising a 'dispensing power' or was pursuing a similar policy to the National League but, in a celebrated phrase, he admitted that the government, 'acting always within the law', had brought pressure to bear on those landlords who refused to follow the example of their more generous peers and act reasonably towards their tenants. He argued, rather unconvincingly, that such a policy was in the best interests of landlord and tenant alike. Echoing Thomas Drummond, the Chief Secretary continued, 'there was no greater foe to the rights of property in the world than the man who attempted harshly to exact its rights and failed to perform its duties'. Hicks Beach was alive to the dangers of the Plan of Campaign and warned that the government would act against it. If the existing law proved inadequate, he threatened to resort to coercive legislation.[19]

The principal effect of such public breast-beating was to infuriate the *Times,* which launched a bitter and sustained editorial attack on the Irish administration, describing its policies as 'scandalous and inexcusable'. The *Times* fulminated:

> There can be no pretence for asserting that the executive power in any part of Ireland is not strong enough to carry into effect the Queen's writ and to place the sheriff in possession of a holding where the tenant is in default. Resident Magistrates and police officers — and even Sir Redvers Buller himself — would be better employed in attending to this elementary duty than in passing judgement upon questions as to the fairness of rents and the solvency of tenants which are out of their proper sphere and in regard to which they can claim no authority. The government can make its appeals to the landlords for generosity, clemency and moderation in the exercise of their legal rights through the regular channels... It is not desirable that officials who have to carry out the business of the courts of law should also be charged with the employment of what is a form of political influence even if exerted for useful public objects.[20]

The nationalists responded with a co-ordinated and well-publicised series of rent collections which displayed a blatant disregard for the law and were a direct provocation to the government.[21] Despite the appeals of landowners and agents alike, the government turned a blind eye to these almost nationwide demonstrations and seemed powerless to act. Its response was hampered by the constraints of the ordinary law. The legality or otherwise of the Plan of Campaign was difficult to determine. It was described in the High Court, in February 1888, by Chief Baron Pallas, as a 'subtle contrivance' which had been devised by a 'master mind'.[22] The government had been advised by its legal officers not to proceed against *United Ireland* for publishing the Plan or anyone else for preaching it.[23] The manifesto itself was unsigned, as were the receipts given to tenants when they lodged their rents with trustees.[24]

The government was convinced that it could secure a verdict against Dillon, O'Brien and a number of the leading campaigners on a charge of conspiracy. This offence, however, could not be punished summarily. The normal procedure was to have depositions taken before a magistrate, who determined whether or not the accused should be sent for trial before a higher court. The government decided to get an authoritative declaration on the criminality of the

Plan of Campaign from the Court of Queen's Bench, the highest criminal court in Ireland, whose judgement was binding on all other tribunals. Once a definite decision was given against the Plan, the government would proceed against the leaders for conspiracy. Dillon had openly advocated the Plan of Campaign at Kenagh, County Longford, on 7 November and at Murroe, County Limerick, a fortnight later, when he delivered what the Chief Secretary termed 'a very violent speech'. It was decided to prosecute him, under an obsolete act of King Edward III, on a charge of intimidation. He was summoned before the Queen's Bench on 26 November 1886 to give bail for 'good behaviour'.[25] Tim Healy, the defence counsel, said he had not been given sufficient time to prepare the case and secured an adjournment.[26]

The events of the next two weeks made the Irish executive even more determined to secure a successful prosecution against the Plan of Campaign and its leaders. On 2 December, the influential liberal newspaper, the *Pall Mall Gazette,* published an interview with the Archbishop of Dublin, William J. Walsh, wherein he defended the agitation on moral grounds.[27] On 4 December, *United Ireland* greatly embarrassed the Irish executive by publishing a confidential 'opinion' of the Attorney-General on the question of rent collections. The Cork Defence Union had urged the Divisional Magistrate, Captain T. Plunkett, to arrest the member for East Cork, W. J. Lane, as he collected rents on the Ponsonby estate. Plunkett refused but agreed to submit their demand to the Attorney-General, Hugh Holmes. *United Ireland* managed to extract a copy of his reply: 'There is a mode by which the landlord might get hold of the money which, of course, is not a matter for the government, and I dare say the landlords will have good advice. I do not see how any action can be taken by the executive'. *United Ireland* crowed: 'The Attorney-General himself blesses the banner of our campaigners'.[28] On the same day, 4 December, Captain Dawson, of Tang, County Westmeath, surrendered to the Plan, the first Irish landlord to do so. Dawson ceded the tenants' full demands; twenty per cent on judicial rents and thirty per cent on non-judicial ones.[29] The *Freeman's Journal* described the victory as 'the most notable event of the week... We have at last discovered one sensible landlord in Ireland'.[30] In an editorial, headed 'First Blood', the leader writer of *United Ireland* claimed, 'the Plan is proved'.[31] The *Freeman's Journal,* whose initial reaction to the agitation had been one of caution, now declared itself unequivocally in favour. 'The Plan of Campaign advances rapidly. A few days ago it was a little driblet. Today it is a stream. A week hence it promises to be an irresistible torrent'.[32] Tim Harrington's brother congratulated him on the success of the Plan. 'You have made history', he wrote.[33] William O'Brien told the Ponsonby tenants: 'We are only in the infancy of this movement but it is rather a promising child... This movement is already an infant Hercules'.[34] An unrepentant John Dillon continued to deliver extremely provocative and intimidatory speeches.[35]

The case against Dillon came to trial in the Queen's Bench on 11 December 1886. Speaking in his own defence, he claimed that he was being victimised and that the law was being manipulated to secure his conviction. The Act of Edward III, under which he was being prosecuted, was devised for use against persons of ill fame, such as robbers, wayfarers, nightwalkers and eavesdroppers, against those who 'mingle ratobane with grain and cast it among fowls, whereby most of the fowls die' and against 'common drunkards and constant haunters of ale houses'.[36] Dillon's argument, that the law was being brought into disrepute, was unavailing and, on 14 December, he was bound over to keep the peace and released on personal bail of £1,000. Of much greater significance was the pronouncement by Justice O'Brien that the Plan of Campaign was 'clearly, distinctly and absolutely illegal'.[37] The way was now clear for the government to take firm action against it.

On 16 December, after a public meeting, four rent offices were opened at Loughrea. The police pounced and confiscated money, books and documents. Dillon, O'Brien, Matt Harris and David Sheehy were arrested on a charge of conspiracy but were immediately released on bail to appear at Loughrea court on Thursday, 23 December.[38]

On 18 December, these four, together with Willie Redmond and Daniel Crilly, were summoned to appear at the Dublin Metropolitan Police Court at 1.00 pm on Thursday 23 December, to answer a charge of conspiracy. It was not until the afternoon of 18 December that the authorities realised that Dillon, O'Brien, Harris and Sheehy were due to appear simultaneously on similar charges in both Loughrea and Dublin. The defendants were then formally notified that the Loughrea prosecutions were being abandoned.[39] The comedy of errors was under way.

At 6.30 pm on the same day, in an attempt to save face, the executive issued a supplement to the *Dublin Gazette* in which the Plan of Campaign was proclaimed to be 'an unlawful and criminal conspiracy'. Henceforth, anyone promoting or joining it was liable to prosecution; money, books, receipts and documents could be seized and those in whose possession they were found could be arrested and brought to trial.

The proclamation, as the *Times* pointed out, did not confer any extra power on the executive, it was merely 'a declaration of purpose . . . a warning'.[40] The government did not have the power to declare the law by proclamation, asserted Tim Healy, and its action was farcical, nonsensical and bordering on pantomime. He additionally maintained that, as yet, no competent tribunal had pronounced on the criminality or otherwise of the Plan of Campaign. Judge O'Brien, in the Queen's Bench, had not made such a pronouncement but had merely delivered an *obiter dictum,* an expression of opinion. He was sitting to try one transaction and, Tim Healy claimed, it was 'mere impertinence' on his behalf to determine and pronounce on another. Healy concluded that the Plan of Campaign was a perfectly legal combination until such time as a judge and jury decided to the contrary.[41]

On the morning of their trial, the *Freeman's Journal* announced that the defendants were expected to perform the feat which was possible only to Sir Boyle Roche's famous bird, that of appearing in two places at once. Although they had been informed that the government was abandoning the Loughrea prosecutions, the defendants chose to present themselves there to surrender their bails, as they were legally entitled to do. The case against them was withdrawn and, simultaneously, in the Dublin Metropolitan Court, the magistrate adjourned the proceedings for a week. The *Freeman's Journal* editorialised scathingly that the 'state prosecutions have all the elements of opera-bouffe'.[42]

When the case was resumed, the defence availed of every opportunity to embarrass the government. The Chief Secretary, the Under Secretary and the Attorney-General were subpoena'd by the defence. Under relentless cross examination, Sir Michael Hicks Beach admitted that a number of landlords had been pressurised to improve relations with their tenants. He declined to comment on the nature of the pressure and refused, repeatedly, to name any of the landlords involved. Sir Redvers Buller acknowledged that he had made representations to a number of Kerry landlords, one of whom was the Marquis of Lansdowne, with regard to rent. Hugh Holmes refused to confirm or deny whether he had written the confidential document published under his name in *United Ireland* on 4 December. Finally, on 11 January 1887, much to the relief of the chief executive officers of the country, the magistrate released the defendants on personal sureties of £500 each and sent the case for trial to a higher court.[43]

The proceedings in the police court portrayed the authorities as inept and indecisive and, according to William O'Brien, made them the 'laughing stock of Dublin'.[44] The manifest incompetence and vacillation shown by the Irish executive provoked all shades of loyalist opinion. Not surprisingly, the *Times* was the most castigatory but even the normally complaisant *Cork Constitution* was censorious.[45] Colonel Saunderson, the leader of the Ulster Unionists in the House of Commons, complained on 13 January: 'I am in despair with the government. They are in a hopeless condition and are letting everything drift'.[46] The Dublin *Daily Express* bluntly asserted that the country was 'swiftly passing into a condition of anarchy'.[47]

Above all, the events of the previous two months had demonstrated the inability of the ordinary law to deal with an agitation like the Plan of Campaign. According to the Chief Secretary, the legal system was cumbersome and unwieldy. It was virtually impossible to get a jury to convict, as the eventual outcome of the conspiracy trials was to show. Sir Michael Hicks Beach suggested a number of improvements, including the extension of summary jurisdiction, the appointment of special juries and a change of venue for criminal trials.[48] The cabinet gave its unanimous assent to these proposals[49] and it was announced in the Queen's speech, on 27 January 1887, that the legal system would shortly be reformed.[50]

During the five weeks that they were on bail, the defendants, in particular Dillon and O'Brien, continued to advocate the Plan of Campaign and to deliver inflammatory speeches.[51] Until they were actually found guilty, the executive had no effectual mean of interfering with their activities.[52] The Plan continued to make progress, its success seemingly inevitable. On 9 January 1887, Lord Dillon yielded on punitive terms. He conceded an all-round abatement of twenty per cent, reinstated the evicted and agreed to pay all legal costs.[53] O'Brien went to Carracastle, County Mayo, to congratulate the tenants:

> We are firing the last volley over the grave of landlordism in Ireland. If they threaten to root out the people like vermin from the land of their fathers I tell them here today that their tottering power will crumble into bankruptcy and nothingness and the grass will grow green around their rent offices . . . Their power will come down like a house of cards before the might of the people.[54]

A few weeks later, Lord Dunsandle, in O'Brien's striking phrase,

CAPITULATION!
LORD DILLON yields to JOHN DILLON and the Plan of Campaign.

took off his coronet to the Plan and surrendered on the same terms as had Lord Dillon.[55]

On 11 February 1887, the Court of Appeal declared Thomas Moroney, a tenant of The O'Grady in County Limerick, a bankrupt. The court decided that the payment by Moroney of a sum of £25.5 as his rent into the 'war chest' and the subsequent sale of his effects under a Plan of Campaign auction were each of them acts of bankruptcy.[56] Hicks Beach had consistently advocated bankruptcy proceedings as the only effective way of dealing with the Plan.[57] This was naive and unrealistic because, as Harrington had pointed out in the Plan of Campaign manifesto, such proceedings were too costly to be generally adopted unless the tenant was the holder of a town park or had some additional industry to the farm in dispute.[58] In addition to renting a thirty-seven acre farm, which had a poor law valuation of £42.25, Moroney held the tolls of a fair, a public house, a forge and five small dwelling houses which had a total valuation of £31. His annual rent was £85 and The O'Grady claimed that Moroney netted from £50 to £60 a year from the tolls of the fair alone.[59] Bankruptcy proceedings may have been an effective legal remedy but, apart from exceptional cases like Moroney, they were too costly to be practicable.

On 14 February 1887, the conspiracy trials began in 'the grimy, inconvenient courthouse', at Green Street, Dublin. They attracted huge publicity. According to the *Freeman's Journal,* 'the press of the three kingdoms and of France and America' was represented. The hackneyed arguments for and against the Plan of Campaign were once more trotted out and finally, on 24 February, after an extremely tedious trial, the jury disagreed and the accused walked free.[60] The humiliation of the Irish administration and the law was complete.

A week after this debacle, Sir Michael Hicks Beach resigned the Irish Chief Secretaryship. He was suffering from physical and mental exhaustion and had developed cataracts of the eyes which threatened blindness. His doctors ordered immediate and complete rest.[61] On 3 March 1887, the diarist and Liberal MP, Alfred Pease, noted that Hicks Beach seemed to be 'depressed, harrassed and in poor health'.[62] In the course of a turbulent sitting of the House of Commons, during which he characterised the Plan of Campaign as 'an organised system of robbery',[63] the unfortunate Hicks Beach was baited relentlessly and was finally provoked into threatening that intimidatory public meetings in Ireland would be broken up by 'something worse than raps with batons'.[64] The nationalists accused him of threatening murder, but Tim Healy refused to be cowed by a 'Mantolini

WEEKLY FREEMAN
NATIONALISTS BEWARE!
BY ORDER M. H. BOTCH
PLAN OF CAMPAIGN
STATE PROSECUTIONS
CAUGHT IN HIS OWN TRAP.
—"Oh! 'Eavings, will not some landlord come and Evict me out of this?"
—"You set the Trap yourself, and, like most of your predecessors, you have got stuck in it, and like them, I'm thinking, there you'll remain."

masquerading as a Cromwell'. Hicks Beach, 'with his ruffled curly hair', recorded Pease, 'did look rather Italian and an agitated one and *was* at this moment rather a pitiful Cromwell'.[65] The wrangling continued unabated until the early hours of the morning and, on the following day, 5 March, Hicks Beach's resignation was publicly announced.[66]

The Ponsonby tenants at Youghal, no doubt convinced that divine providence was exacting retribution for Hanlon's death,[67] expressed the feelings of nationalist Ireland when they hailed the Chief Secretary's resignation 'as an assurance of the speedy and utter overthrow of the present Tory government of botches and coercionists and of the inevitable triumph of our cause'.[68] The *Freeman's Journal* asserted bluntly that it was better for Hicks Beach to retire than to persevere with policies of 'folly and blundering' and concluded, disdainfully: 'A cold in the head is seldom pleasant and diseases of the eye are often troublesome but assuredly he is well rid of the Chief Secretaryship at the cost of influenza and incipient cataracts'.[69]

Given the circumstances of his resignation, it was inevitable that the tributes paid to Hicks Beach in England should have been fulsome.[70] The Prime Minister was warm and generous in his praise and said that Hicks Beach would be a great loss not alone to the government but to the country generally. However, having praised his quondam Irish Chief Secretary, Salisbury proceeded, by implication at least, to bury him. He compared Ireland to a bad dream which could not be dissipated, a nightmare from which one could not awaken. He maintained that the government had been too soft towards Ireland and the policy of 'platitude and rosewater' had clearly failed. The Prime Minister recognised the necessity for remedial legislation but stressed that the law must first be strengthened and enforced.[71] Privately, he expressed his sentiments a good deal more crudely: The Irish 'must "take a licking" before conciliation will do them any good'.[72] Fortunately for Salisbury, Hicks Beach's successor was a man of sterner resolve.

Hicks Beach's term as Irish Chief Secretary may be summarised, in a phrase used to describe Buller's Kerry campaign, as 'a sort of mixture of Charles Lever and Dion Boucicault'[73] and earned for him the unfortunate sobriquet, 'Mickey-the-Botch'. His forbidding persona and quixotic policies alienated the loyalist population and generally demoralised the Irish administration.[74] The Chief Secretary's leadership was typified by the bungled attempts of the executive to confront the Plan of Campaign and prosecute its leadership. But the executive was constrained by the inadequacy of the existing legal

ANOTHER WRONG MAN GONE RIGHT.

Mickey the Botch.—"They knew the end so well that they had my name already engraved there——Poor Balfour!!!"

system and, as the government was unwilling to hold a parliamentary session in the autumn of 1886, the law could not be amended until the spring of 1887, at the earliest. The government's lack of a clearly defined Irish policy, a vocal but ineffectual Prime Minister and a cabinet rent by internal dissension made Hicks Beach's task in Ireland an unenviable one. As Chief Secretary, he displayed a genuine sympathy for the Irish tenants and refused to administer Ireland on a pro-landlord basis. As Tim Healy noted, Sir Michael Hicks Beach's 'icy exterior concealed an honest heart'.[75]

5 *'Tiger Lily'*

This young gentleman has three qualifications for the post. He is the nephew of Lord Salisbury; he has no reputation for statesmanship to injure, and he knows nothing of Ireland.
Freeman's Journal, 7 March 1887.

I took up the other day... *United Ireland* and I find in the leading article that it called Mr Balfour, "Bloody Balfour" six times. It has invented many epithets for him. It has called him by the name of almost every animal in the zoological gardens; and having exhausted natural history they are now calling him "sleuth tiger", an animal I never heard of before but which may exist in some parts of Ireland I have never visited.
Reginald Lucas, *Colonel Saunderson,* MP, p.135.

Sir Michael Hicks Beach, along with the liberal and nationalist press and opposition politicians in general, had little faith in the ability of his successor to survive the rigours of the Irish Chief Secretaryship. On the day he resigned, Beach warned the Prime Minister that, if Arthur Balfour's health failed in the wake of his own collapse and the sufferings endured by Forster and Trevelyn, the Home Rule argument would be greatly strengthened. In an obvious reference to Balfour's spartan frame and notoriously languid manner, he claimed that, in the Irish office, physique was quite as important as ability and urged that Balfour, if he had to be appointed, should be discouraged from trying to do too much.[1]

Balfour's task in Ireland was indeed immense. The administration was demoralised by the inadequacy of the legal system, the Plan of Campaign was deeply entrenched and appeared impregnable, landlords were disorganised, while tenants were disaffected and alienated and the countryside, especially in the west and south-west, was seriously disturbed. In February 1887, J. Byrne reported that the western division was in 'a state closely bordering on rebellion'. Two months later, he noted that there was 'a very dangerous spirit throughout the division'.[2] In March, Colonel Alfred Turner reported that Counties Kerry and Clare were 'in a lawless and most unsatisfactory state' and that the demeanour of the people was 'sullen and defiant'.[3] In the first quarter of 1887, agrarian crime increased by more than forty per cent over the previous three months. Agrarian murders were committed in Kerry, Clare, Louth and the East Riding of Galway.[4]

The difficulties facing Balfour were neatly demonstrated by the public reaction to the arrest and imprisonment of Canon Keller of Youghal, leader of the Plan of Campaign on the Ponsonby estate, for refusing to disclose the names of the trustees of the tenants' 'war chest'. Following so shortly on the fatal police bayonetting of Hanlon, a Youghal fisherman, Keller's imprisonment outraged nationalist Ireland. Alfred Pease recorded a digest of the proceedings as reported in the *Times:*

> A scene of indescribable tumult and disorder followed Judge Boyd's order of committal. The public in the court cheered the prisoner, Archbishop Walsh accompanied him to the gaol, the horse was taken out of the cab, the crowd drew him to Kilmainham, a dozen policemen followed, then the Lord Mayor of Dublin's coach, aldermen and MPs, then six carloads of police, then even more cars and cabs full of sympathisers. Opposite the spot where Emmet was hanged the procession halts, uncovers and shouts, "God Save Ireland", cheers for the Plan of Campaign and groans for Judge Boyd. Father Keller enters the prison arm in arm with the archbishop; he is received there by the president of Maynooth College (Rev. Dr Brown) while William O'Brien MP harangues the crowd, envies Keller his honours and his "sufferings for Ireland", denounces alien rule, and appeals to it to follow in Father Keller's footsteps. And Balfour (Balfour, mind you!) is going to stamp all this out.[5]

To the surprise of most and the consternation of many, Arthur Balfour proved to be, as Tim Healy expressed it, 'the most remarkable of Irish Secretaries'.[6] On 28 March 1887, three weeks after taking office, he introduced the long-awaited crimes bill into the House of Commons and, a few days later, a land bill was placed before the Lords. Both measures were highly contentious and, despite repeated use of the guillotine to shorten the debate, the crimes bill did not become law until 19 July; another month was to elapse before the land act came into force.

During the four months that the polemic on the government's 'kicks and ha'pence' policy raged in the legislature, evictions were resumed in Ireland on a large scale. Sir Michael Hicks Beach had actively discouraged evictions and, during his Chief Secretaryship, not a single tenant had been ejected from any Plan of Campaign estate. However, within days of Balfour's appointment, evictions commenced on the Shirley property in County Monaghan. Between mid-March and the end of May 1887, 7 families were evicted from C. W. T. Ponsonby's estate in Cork, 10 from P. J. B. Daly's in Mayo, 22 from the Wynne property in Leitrim and 55 from the Annesley estate in Cavan. More than 200 individuals lost their homes on Lord Granard's County Longford property, while Denis Kilbride and some 300 other individuals were evicted by the Marquis of Lansdowne in Queen's County.[7] However, it was the eviction campaign on Colonel John O'Callaghan's Bodyke property in County Clare, during the first half of June, that enraged the nationalist and liberal population of Ireland and Britain.

O'Callaghan was tortuously described by *United Ireland* as a man with 'a maw capacious as a shark's, a heart hard as a nether millstone and a plentiful lack of the bowels of compassion'.[8] The Divisional Commissioner, Colonel Alfred Turner, a man not generally known for his sensitivity, castigated him, in January 1887, as a 'harsh exacting landlord', who was very much alive to the rights of property but blind to its duties and whose relations with his tenants had, for more than thirty years, been 'as bad as they could be'. Turner reported that O'Callaghan was 'bent upon evicting, utterly regardless of humanity and of the peace and welfare of the district and country'. The whole county, he continued, was apprehensive and he feared that O'Callaghan's tenants would be aided by thousands of men from other localities. Colonel Turner concluded that, if the government were to aid O'Callaghan by lending him crown forces, serious crimes and disturbances would ensue, which would have 'the worst possible effect on public opinion' in England and which would place a very potent weapon in the hands of the 'agitators'.[9] After Sir Redvers Buller and Colonel Turner had both unsuccessfully appealed to O'Callaghan to abandon the evictions, the Dublin administration forced them to be postponed on the grounds that they were 'indefensible'.[10]

On 26 May 1887, the *Freeman's Journal* noted that a force of 200 soldiers and between 300 and 400 policemen had been drafted into Bodyke to evict 'some thirty-two starving families'. Colonel Turner reported that 'the whole of County Clare is more or less in a state of excitement' and again complained that government forces were being badly employed in aiding 'an unreasonable, harsh and vacillating landlord', whose conduct had, for years past, led to much of the disturbance in east Clare.[11]

In the event, he seriously underestimated the impact of the Bodyke evictions. They lasted almost two weeks and secured a notoriety unsurpassed during the whole era of the Plan of Campaign. Some forty-three people, most of them women, were arrested on charges of assault and obstruction. The evictions and their attendant scenes were extensively and emotionally portrayed in the nationalist and liberal press and provoked anguish and anger in Ireland and Britain.[12]

An infuriated and despairing Michael Davitt, addressing the assembled tenants at the close of the first day's evictions, urged all the young men to take an oath similar to one he had taken thirty years previously, 'to bear towards England and England's government in Ireland all the concentrated hatred of my Irish nature'. He deplored the lack of resistance shown by those who had been evicted but blamed it on the advice he and others had given to the tenants over the past number of years, to obey the law and forego violence and outrage. 'I am heartily ashamed of ever having given such advice to the Irish people... and I make a resolution here today that as long as this work continues not a word of censure will fall from my lips against an Irishman who defends his home at the risk of his life'. He also regretted that they did not have the weapons to teach the evictors a lesson they would never forget. Davitt continued:

> Would to God we had the means, the weapons by which freemen in America and elsewhere have struck down tyranny... but here we are first disarmed by this cowardly assassin government of England, deprived of weapons and then when they have made us thus defenceless they send armed cohorts into these peaceful valleys and along the mountain sides to cut your throats and spoil your homesteads.

Davitt swore to continue the struggle until 'landlord tyranny and English government' had been destroyed in Ireland.[13]

At a National League meeting in Dublin on 7 June, Tim Harrington condemned the advice given by Davitt and, on behalf of the organising committee of the League, urged the tenants to maintain their self-control and not to resort to violence which, he said, would merely be playing into the hands of 'the present coercion government'.[14] After further turbulent eviction scenes on the following day, an unrepentant Davitt reiterated what he had said previously: 'He was heartily ashamed that he had ever counselled the extreme moderation on the part of the Irish race and made a resolution never to do it any more'.

He urged the people of Bodyke 'to give all reasonable resistance to those armed burglars'.[15] The *Freeman's Journal* protested in the strongest possible terms against the government's 'insane policy' and condemned the evictions in Bodyke 'as a scandal to the administration and a formidable danger to the peace and order of the country'.[16] The *Pall Mall Gazette* responded melodramatically:

> If the whole English people could have been condensed into one eye, with one brain behind it to understand and one heart to feel and could have looked for a moment down that great slope near Bodyke the Irish question would have been settled before another sun set on two estranged and embittered peoples.[17]

John Dillon moved the adjournment of the House of Commons on the question of Irish evictions and a particularly acrimonious debate ensued.[18] In the face of mounting public disquiet, Balfour called on Buller to consider the possibility of deferring evictions or, at any rate, those in which landlords were acting harshly, until after the land bill had become law.[19]

But the landlords were determined to press ahead with ejectments before the eviction clauses of the government's bill could come into force. More than seventy families were evicted from the Brooke estate in County Wexford in a campaign which started on 6 July and lasted a fortnight.[20] Davitt was present and made a scathing attack on the tenants' tactics. He condemned the futility of 'cheering Fr Dillon, groaning Hamilton and calling the emergencymen names'. He criticised, as he had done at Bodyke on 8 June, the attitude of the organising committee of the Irish National League which be believed to be too moderate and too conservative.[21] John Redmond described Davitt's advice to the Wexford tenants as 'little short of folly' and he, Dillon and O'Brien deprecated Davitt's attitude to the organising committee.[22] The Archbishop of Dublin, William J. Walsh, who had taken a special interest in the struggle on the Brooke estate, was prompted to write to the *Freeman's Journal* urging the government to introduce, and speedily pass, a short bill suspending the power of eviction in Ireland until after the land bill had become law.[23] Practically, the government was endeavouring to implement just such a policy, in accordance with Balfour's instructions of mid-June, but was meeting stiff resistance from landlords, agents, sheriffs and sub-sheriffs alike. Colonel Alfred Turner reported to Buller at Dublin Castle, towards the end of July: 'I understand my line to be to postpone every eviction I possibly can 'til after the passing of the land bill'. While he had secured some deferments, there was 'a perfect sheaf of applications' from Clare and he found himself 'very heavily pressed to carry out evictions *all round'*. Turner blamed the sub-sheriffs whom, he claimed, feared the loss of their fees should the eviction clauses of the land bill come into force.[24] The climax was reached in mid-August, days before the enactment of the bill, when William O'Brien and John Mandeville forced the Countess of Kingston to postpone evictions on her Mitchelstown estate. However, in inciting the tenants to resist evictions, O'Brien and Mandeville had violated the provisions of the new Crimes Act and the government determined to prosecute.

The Criminal Law and Procedure (Ireland) Act, 1887, known as the Jubilee Coercion Act or more commonly as the Crimes Act, gave far-reaching powers to the government and greatly strengthened its hand in dealing with agrarian agitation. The first section enabled resident magistrates to question, privately and on oath, any individual they considered capable of providing information. Under the second section, they were empowered to deal summarily with the following offences: conspiracies against the payment of rent, boycotting, intimidation, unlawful assembly, resistance to eviction and the inciting of others to commit any of these crimes. The maximum penalty for each offence was six months imprisonment with hard labour, with the right to appeal in each case. These two sections were primarily intended to deal with the Plan of Campaign. Under the remaining provisions of the act, the authorities could order a change of venue in jury trials—to guarantee the impartiality of such trials, they claimed; to secure convictions, retorted the nationalists—the Lord Lieutenant could declare by proclamation that these provisions were in force in any specified part of the country, known as proclaimed districts, and, in addition, he could suppress any organisation engaged in illegal activities.[25]

The Chief Secretary was well aware that purchase was the only panacea for the land question but had to postpone that solution to another session. Although Balfour declared that the Land Act of 1887 was merely a temporary expedient, it was nonetheless a constructive measure and, by admitting leaseholders to the benefits of the 1881 Land Act, conceded a long-standing nationalist demand. The land courts were empowered to revise all judicial rents fixed between 1881 and 1886. The new judicial rents were to be based on prevailing prices and labour conditions in each district and were valid for three years. A county court judge could stay the eviction of a tenant whose insolvency was not his own fault and could spread the payment of arrears over a reasonable period. An 'eviction made easy' clause simplified the whole eviction process.[26]

Many members of the Tory party, not least the Prime Minister, were unhappy at these provisions. While recognising the necessity for the land act, Salisbury admitted that it caused him 'pain and grief'.[27] Many landlords felt betrayed. The Marquis of Waterford told Lord Ashbourne, in November 1887, that the Plan of Campaign had just been adopted on his own estate and that the new land measure was directly responsible. 'I never expected to be ruined', he complained bitterly, 'by the party that my family has spent hundreds of thousands to support'.[28] William O'Brien later claimed on a number of occasions that the Land Act of 1887 was the 'direct offspring of the Plan of Campaign'.[29] The act, by ensuring the revision of judicial rents, which the Tories had always regarded as sacrosanct, and by admitting more than 100,000 judicial tenants to the benefits of the 1881 Land Act, was the single most important achievement of the Plan of Campaign.

On the other hand, the Crimes Act proved to be a turning point in the history of the agitation. Henceforth, the initiative lay with the government. Section one inquiries, or star chamber courts as they were commonly called, were held into the operation of the Plan of Campaign on specific estates, such as Wybrants Olphert's in Donegal and Lord De Freyne's in Roscommon. Certain localities, where the Plan was deeply entrenched, such as Woodford, County Galway; Balla, County Mayo; Mullinahone, County Tipperary; Arthurstown, County Wexford; and Castlerea, County Roscommon; were also investigated. Several prosecutions resulted and in all cases information inimical to the Plan was elicited.[30]

On 6 January 1888, James McLaughlin from Cloonfad, County Roscommon, was badly beaten by six disguised assailants while on his way to Ballinagar fair. His pigs were scarred, rendering them unsaleable. McLaughlin had refused to join the Plan of Campaign on the Worthington estate.

On 9 February, a Special Branch officer named Gibson suggested to the Attorney-General, Peter O'Brien, that a sworn inquiry, under section one of the Crimes Act, into the assault on McLaughlin would 'open up the whole Plan of Campaign conspiracy at Castlerea'. The police already had a good deal of reliable information and could produce several witnesses who had joined the Plan and who would probably confess all they knew under oath. 'The mere opening of such an inquiry if promptly done would probably break the whole conspiracy', he concluded. On 22 February, District Inspector

Meredith made a sworn statement that the district of Castlerea had been proclaimed under the Crimes Act and that the outrage on McLaughlin had been perpetrated to intimidate him into joining the Plan of Campaign. On 14 March, the local Resident Magistrate, N. L. Townsend, was instructed by the Attorney-General to begin a sworn inquiry into the case.

The investigation unearthed the entire saga of the Plan of Campaign on the estates of H. S. Packenham Mahon, J. C. Murphy, Mr Worthington and G. W. Sandford and produced sufficient evidence to implicate more than forty individuals in the conspiracy, including John Fitzgibbon, 'the chief promoter and worker of the Plan in the Castlerea neighbourhood' and a number of MPs, including J. R. Cox, Conway, Willie Redmond, Daniel Crilly, David Sheehy and John Dillon. It was decided not to proceed against the elected representatives but Fitzgibbon and six locals were charged with taking part in the Plan of Campaign and of inciting tenants not to pay their rents.[31]

Divisional Commissioner Byrne noted with satisfaction that the imprisonment of Fitzgibbon, whom he termed 'a most mischievous and dangerous agitator', was expected to have 'a very beneficial effect'.[32] It was certainly a blow to the organisation of the Plan of Campaign throughout County Roscommon. At the beginning of September 1888, the Plan came to an end on the Packenham Mahon estate. The tenants had originally demanded an all round reduction of forty per cent but settled for abatements ranging from fifteen to twenty per cent, with the landlord paying all the costs. The tenants demanded their campaign rents from Fitzgibbon, who was the treasurer of the 'war chest'. According to the police report, he paid over the rents but said, 'it was the last time he would ever try to do anything for them and that he was being betrayed all round'.[33]

William O'Brien and John Mandeville were the first individuals to be prosecuted under the Crimes Act of 1887. On 11 and 13 August, they both urged the Kingston tenants to resist evictions, an offence punishable under section two. After their trial at Mitchelstown on 9 September, which neither attended, three civilians were fatally wounded by the police. This event was emotionally, if hyperbolically, designated 'the Mitchelstown Massacre' and earned for the Chief Secretary the unenviable sobriquet of 'Bloody Balfour'. Gladstone's catch-phrase, 'Remember Mitchelstown', came to symbolise the union of hearts that bound nationalist Ireland to liberal England.[34]

The Chief Secretary's resolute defence of the Mitchelstown police and magistrates in the face of almost universal public criticism had an incalculable effect on the morale and confidence of the constabulary and castle officials alike.[35] During the ensuing years, the Crimes Act of 1887 was relentlessly implemented by a revitalised Irish administration, inspired by the purposeful leadership of Arthur Balfour. Up to 7 May 1889, twenty five MPs, including C. A. V. Coneybeare, member for North-West Cornwall, were sentenced to prison terms for offences committed under the act. Another notable 'martyr' was the English eccentric and Home Rule sympathiser, Wilfrid Scawen Blunt. 'I was delighted to see you had run Wilfrid Blunt in', Lord Salisbury complimented his nephew in October 1887.[36] John Dillon, David Sheehy and James Gilhooly were each gaoled on more than one occasion, while William O'Brien was sentenced to no less than six terms of imprisonment between 1886 and 1891.[37]

Most of the transgressions were connected with the agrarian agitation, such as inciting to join the Plan of Campaign or to boycott or to resist evictions. David Sheehy was prosecuted for advising tenants at Frenchpark on 25 September 1887 to keep the roofs over their heads: 'If Lord De Freyne makes an attempt to evict you I hope there is not a man, woman or child but will barricade their holdings and will assist each other against the evicting forces'. After serving two separate prison terms for inciting to boycott, James Gilhooly, MP for west Cork, said, 'he would be glad to take part in the Plan of Campaign whenever he may be called on to do so'. W. J. Lane, MP for east Cork, said, in mid-January 1888, that he had 'advocated the Plan of Campaign and, please God, I will do it again. I will do it before I go into gaol and after I come out, if there is a necessity for it'. Alexander Blaine was sentenced to four months imprisonment, which was increased to six months on appeal, for advocating the Plan at Derryart, County Donegal, in early January 1888. 'Mark the enemy and shun him', he advised. 'The word used to be in the past, agitate!, agitate!, agitate! The word today is boycott!, boycott!, boycott!'[38]

Several Plan of Campaign priests were imprisoned under the Crimes Act. No doubt they drew inspiration from the guidance given to Canon Keller of Youghal by the Bishop of Cloyne. Dr John McCarthy intimated his approval of Keller's activities on behalf of the Ponsonby tenants and in particular his refusal to disclose the whereabouts of the tenants' 'war chest'. Dr McCarthy counselled:

> There is something above law and that is justice and while circumstances may arise...that would warrant one in disobeying the law, no circumstances can ever arise that would justify a man in acting in opposition to the plain dictates of justice and it is by these

> latter precepts that I am sure that any course you may adopt will be guided.[39]

Fr Matthew Ryan of Herbertstown was sentenced for inciting tenants to join the Plan at Caherconlish, County Limerick, in December 1887. 'The lease of landlordism is expired', he said, 'or will expire in a few months if you are only men enough to adopt the Plan of Campaign'. The Donegal priests, McFadden and Stephens, served the maximum sentences on a similar charge. Fr John Maher of Luggacurran warned Lord Lansdowne's agent, in March 1889, that 'he might as well bay the moon or try to stop the tide with a broom as break up the Plan of Campaign in Luggacurran'. Canon Doyle of Ramsgrange, County Wexford, was sentenced for proposing the following resolution in September 1887: 'That boycotting, as it is called, when prudently practised, is the most effective means of crushing land and grass grabbers and to bringing scoundrels, ruffians and renegades to their senses'.[40]

The Prime Minister stated, at the Guildhall in London on 9 November 1887, that a marked improvement was already noticeable in Ireland, even though the Crimes Act had only been in operation for a few months, and he promised that the government would persevere in its efforts 'to restore the reign of law and order in Ireland'.[41] A week later, Sir Redvers Buller complimented Sir Joseph West Ridgeway, his successor as Under Secretary at Dublin Castle: 'It seems true that you are winning in Ireland all along the line'.[42] The *Times* regarded the imprisonment of William O'Brien as 'a turning point in the long and arduous struggle between the forces of anarchy and the forces of the law'.[43] In their reports for November 1887, Divisional Commissioners Byrne and Turner commented on the beneficial impact of the Crimes Act, the latter, as usual, in his avuncular manner:

> The Crimes Act has already produced a great effect and if the fountain head is boldly attacked the success of it will be complete... It is no hard task to govern Irishmen. A firm, steady hand is all that is required and a few months more of the present system of administration will restore law, order and peace and consign to their proper place the vile agitators who can no more exist in a healthy state of society than can the worms which prey upon corruption in a sound, healthy and living body.[44]

County Court Judge R. Ferguson, after completing 'a very prolonged session' in West Cork, informed the Attorney-General that the effect of the recent legislation was 'admirable'.[45] Early in the new year, the Chief Secretary noted:

> On the whole, the general tendency of events seems satisfactory... It is said that rents are being better paid, that a better feeling prevails, and that intimidation is greatly losing its effect in many parts of Ireland... The prosecutions under the Crimes Act have, on the whole, been very successful and the local tyrants are just beginning to find out that prison discipline and the plank bed are after all not so agreeable as they were disposed to think while the Crimes Act was passing... The result is satisfactory.[46]

It is more difficult to assess the specific impact of the Crimes Act on the Plan of Campaign. Section one inquiries undoubtedly weakened the agitation at local level. On the other hand, the prosecution, some would say persecution, of politicians and priests more than countered this effect. The prison treatment meted out to political prisoners, the tragi-comedy of 'O'Brien's breeches' and the escapades of individuals like Charley Tanner and Jasper Pyne—both of whom successfully evaded arrest for several months—produced by turn outrage and laughter, which encouraged tenants to play their part in the struggle by upholding the principles of the Plan of Campaign. Tenants were urged, and frequently resolved, not to pay rent until the political prisoners, lay and clerical alike, had been unconditionally released. On 13 November 1887, James Gilhooly MP urged an assembly of farmers to refuse to pay

> one farthing of rent to their landlords until they who were the practical gaolers of William O'Brien should tell him to go forth a free man. The tenant farmers of Ireland would be unworthy of the name of men if, by their action in refusing to pay rent, they did not avenge the outrage and the cruelty practised on this fearless and spotless patriot.[47]

On New Year's Day 1888, the following notices were extensively posted throughout Limerick and Tipperary:

> Tenant farmers of Munster and of Ireland in general, but particularly of Lattin:
> Now that the Tory government has thrown down the gauntlet and has incarcerated the pure-souled general, Father Mat Ryan, and others of our noble leaders, you are hereby commanded not to pay a single stiver of rent until they are all unconditionally released. Stand firm to your guns. If not you may soon look out for squalls.
>
> By Order, God Save Ireland.[48]

6 *The Pope One Day, Parnell the Next*

> The Irish agitation in its various forms is destroying the moral forces that hold society together. The conscience of the people is becoming perverted. Civil government can with difficulty maintain itself under such conditions... Every student of history must recognise that the present state of things in Ireland is very similar to that which prevailed in countries where Revolution, Atheism and Anarchy followed in the worst forms. The Pope, of all others, so frequently attacked by the spread of ideas which are subversive to all authority, could not fail to see this and take whatever measures he was able to employ to check the current which has unhappily reached Ireland.
> Colonel John Ross of Bladensburg, 'Notes on the action of the Pope in Ireland', 10 March 1887 (BL, Balfour Papers, Add. Ms.49821).

> This movement, as I am sure you understand, is to a great extent a game of bluff on both sides.
> John Dillon to Charles Stewart Parnell, 14 January 1889 (TCD Ms.6745, John Dillon Papers).

On 8 April 1888, public meetings were held at Loughrea, County Galway, and Tullyallen, County Louth, at Macroom and Kanturk in County Cork and at Ennis, Kilrush, Scariff and Miltown-Malbay in County Clare, in answer to the Chief Secretary's recent assertion that the National League was a thing of the past in the proclaimed districts. The Ennis, Loughrea and Kilrush meetings were dispersed by force, seventy-four people were arrested at Ennis and twenty-three, including William O'Brien, at Loughrea. John Dillon was prosecuted for taking part in a criminal conspiracy and for promoting the Plan of Campaign on Lord Massereene's County Louth estate.[1]

O'Brien was accompanied to his trial at Loughrea on 26 April by Wilfrid Blunt, who noted in his diary that both O'Brien and the Bishop of Clonfert, Dr. Patrick Duggan, were apprehensive 'with regard to Rome'. Rumours were circulating wildly that a papal rescript, condemning the Plan of Campaign and boycotting, was imminent. On the following day, the substance of the Pope's directive was published in the *Freeman's Journal*. 'I watched O'Brien's face as he read it', recorded Blunt, 'and it grew paler and more determined... The news lies like an incubus on us all'.[2]

The circular which was issued on 20 April, under the signature of R. Cardinal Monaco, condemned the Plan of Campaign on three grounds: it was unlawful to break contracts freely entered into by landlord and tenant; the land courts were available to those who believed their rents were not fair; the monies collected under the Plan had been extorted from tenants. Boycotting was denounced as being contrary to the principles of justice and charity and censured as an instrument of persecution used against individuals willing to pay their rent and against those exercising their right to take vacant farms.[3]

On the day that details of the rescript appeared in the Irish papers, Dillon addressed The O'Grady's tenants at Herbertstown, County Limerick. He claimed that there was no such thing as free contract between landlord and tenant in Ireland and that it was absurd to insist that it was immoral to break non-existent contracts. The land courts were strongly biased in favour of landlords and, additionally, lease-holders had been excluded from them until the Plan of Campaign agitation had forced the government to change the law in 1887. Dillon denied that tenants had ever been compelled or intimidated into joining the Plan. He said he would never countenance such methods.[4]

It was not until lunchtime on the following day, 28 April, that the Bishop of Clonfert formally received his copy of the rescript. O'Brien, his counsel and some sympathisers, including Wilfrid Blunt, were gathered at Dr Duggan's hospitable table when a letter, bearing 'a foreign superscription', was delivered. The bishop gave 'an extra-ordinary chuckle and a long whistle' and then read out the contents and each laboriously-deciphered paragraph was greeted with roars of laughter. Dr Duggan contemptuously dismissed the rescript by instructing his manservant, Mike, to 'kill another pig'. Like Dillon, the bishop and his guests had seen the flaw in the Pope's circular. 'It is clear', Blunt noted, 'that the document can be evaded and will probably prove a dead letter, for it is founded on a misapprehension of facts'.[5]

On the same day, Archbishop Croke discussed the rescript with Dillon. The result of their colloquy was an article in the press on 1 May which was almost jesuitical in its ingenuity. It was argued that the Plan of Campaign should indeed be condemned if it fell within the

THE NEW PAPAL NUNCIO.

Balfour, who has got on a suit which is kept in the Castle for Chief Secretaries to wear on certain occasions, to Erin.—"Ha! Ha! See what I have got over from Rome. Now, I think you will give me up that paper?"

Erin.—**NEVER**!!! That document in your hand was obtained by fraud and misrepresentation, and has therefore no moral force."

Balfour.—"Ho!! Ho!! Then you mean to split with Rome?"

Erin.—"**NEVER**!!! I have always been a faithful child of the Church, and will ever remain so, but I mean to manage my own domestic affairs, as everyone else does."

categories outlined in the circular but, as the rescript itself was based on an incorrect assumption of fact—most land tenure in Ireland was not based on a simple relationship between letter and holder but on dual ownership, established by the Land Act of 1881— it was not applicable to Irish conditions.[6]

On 4 May, at Kilmurray, County Clare, Dillon said that the rescript was not binding on the conscience of Irishmen.[7] Three days later, he used the occasion of being made a freeman of the borough of Drogheda to deliver a major speech. He asserted his allegiance to the Church in spiritual matters but condemned Rome's interference in Ireland's political affairs. With Swiftian overtones, he demanded:

> Are we to be freemen in Ireland or are we to conduct our public affairs at the bidding of any man who lives outside Ireland? We owe it to ourselves, we owe it to our friends in England, we owe it to the ancient traditions of our country, we owe it to our Protestant fellow countrymen who expect they are about to share with us a free Ireland that it will not be an Ireland that will conduct its affairs at the bidding of any body of cardinals.

He concluded that if the freedom of Ireland meant substituting Rome for Westminster he would cease at once and forever to struggle for Irish independence.[8] Dillon was at once orchestrating and reflecting the temper of the Irish people generally. George Bernard Shaw, in an unsigned leader in the *Star* on 30 April, wrote: 'It is time for the people of Ireland to declare that they accept advice and counsel in their political concerns from no power on earth but the political leaders they themselves have chosen'.[9] An unusually trenchant *Freeman's Journal* editorialised on 11 May:

> We uphold in the most strenuous manner the authoritative dictum that in matters of politics the opinion and judgement of the Irish people on Irish affairs must be supreme. Nothing shall shake the people or their leaders or make us waver in the undaunted assertion of that inalienable right against all comers. It is the cornerstone of the freedom of our people. It must remain as fixed as the rock.[10]

On the following day, 8 May, Parnell addressed a meeting of prominent liberals at the Eighty Club in London. In his speech, which the *Freeman's Journal* described as 'magnificent', he dealt with the rescript and, at much greater length, with the Plan of Campaign. Effortlessly, almost contemptuously, he dismissed the rescript as bound to be 'a disastrous failure' and said it was not his business to interfere with the line his Catholic colleagues would take 'in vindication of their political freedom against the claim which has been put forward from this distant country in reference to the affairs of Ireland'.[11]

When the Catholic members of the parliamentary party duly met in the Mansion House, Dublin, on 17 May, they passed a series of resolutions rejecting the allegations contained in the rescript. They acknowledged the spiritual jurisdiction of Rome but asserted that 'Irish Catholics can recognise no right in the Holy See to interfere with the Irish people in the management of their political affairs'.[12]

Three days later, the citizens of Dublin voiced their protest by means of a popular demonstration in the Phoenix Park. The Lord Mayor, Thomas Sexton MP, set the tone of the proceedings:

> We Irish Catholics are connected... with two powers which are situated beyond the shores of Ireland. By our free consent we are linked in our spiritual affairs with the Holy See. Against our will we are linked for the purpose of secular laws to the will of the parliament of England. Our religion is independent of England and our politics are independent of Rome.

John Dillon, William O'Brien, John Redmond, Daniel Crilly, Dr Kenny and J. R. Cox, all of whom were involved with the Plan of Campaign, expressed similar sentiments. The citizens of the capital endorsed the resolutions which the Catholic MPs had adopted at the Mansion House meeting.[13]

Similar public meetings were scheduled to take place at other centres on the following Sunday, 27 May. A lone, dissenting, though not entirely unexpected, voice was now raised. Dr O'Dwyer had stated in a letter to the Mayor of Limerick that the rescript was binding on the conscience of each Catholic and, while it was a grievous sin to disobey, it was an even greater one to deny the Pope's authority to issue the circular.[14]

Dr O'Dwyer was publicly and roundly condemned. Speaking in Limerick, the bishop's own diocesan centre, William O'Brien called him 'an arrogant ecclesiastical official'. At Kildare, John Dillon denied the right of Dr O'Dwyer to speak on behalf of the Catholic Church in Ireland, insisting that the majority of Irish bishops supported the Plan of Campaign and that Dr O'Dwyer spoke for himself alone. Dillon cast him as

> the champion of the landlord interest in Ireland...He is the supporter and the champion of rack-renting and landgrabbing in Ireland...He represents a small minority of the Roman Catholic

> Church in Ireland who are on the side of the oppressor against the oppressed, of the rich man against the poor man, of the landlord against the tenant.[15]

Throughout the country, successive speakers drove home the point that Rome had no right to interfere in Irish politics. At Wexford, John Redmond insisted that 'if Home Rule were to mean Rome Rule, he for one would not be a Home Ruler'. Davitt, speaking at Bray, warned that nationalists would not 'exchange the servitude of Westminster for that of the inquisition, nor the coercion of Mr Balfour for that of Dr O'Dwyer', a sentiment with which the leader writer of the leading nationalist daily wholeheartedly concurred: 'Irishmen would die rather than desert their faith at the bidding of England or any other power. But Irishmen would die too rather than desert their country's cause at the bidding of Rome. In establishing this the rescript or decree has done a splendid service'.[16]

The most emphatic refutation of the papal rescript was delivered by William O'Brien at Glensharrold, in the diocese of Limerick, on 29 May. 'We will stick to the Plan of Campaign', he pledged. 'We will stick to boycotting and will treat every landgrabber and every exterminator as a public pest and a public enemy more dangerous to society than a leper is'.[17]

On the following day, the Irish hierarchy met at Clonliffe College and, in a series of vaguely-worded resolutions, effectively endorsed the position taken by the parliamentary party. The rescript, as interpreted by the bishops, was intended to instruct the Irish people in the sphere of morals; it was not an attempt to interfere in Irish politics.[18] The only member of the hierarchy to publicly break ranks was the unrelenting Dr O'Dwyer. At a diocesan synod on 11 June, he repeated the points he made in his letter to the Mayor of Limerick a fortnight earlier. O'Dwyer insisted that the rescript was binding on both the clergy and laity of his diocese and said that those who agitated against it were committing 'heinous offences'.[19]

With the exception of the Limerick diocese, the papal rescript had little or no effect on the agrarian agitation. In general, at local level, the clergy who had pledged themselves to the Plan of Campaign continued to support it. The Plan was adopted on at least seven properties after the rescript was issued[20] and the Special Branch noted that on one of these, the Eyre estate in County Tipperary, the local priests supported the tenants and encouraged the agitation.[21] At the end of June 1888, Colonel Alfred Turner reported from Ennis that the Clare priests, despite the decree, were giving 'great trouble'.

Six months later, he opined that the majority of priests in Clare and Kerry had treated the rescript 'with open contempt'.[22] In March 1889, Colonel John Ross of Bladensburg, the prominent Catholic Unionist who had served as secretary to the British mission to the Holy See in 1887, noted that the rescript was a dead letter. 'The clergy support the agitation as vigorously as ever', he wrote.[23] An allegation in the *Irish Times* that the Catholic clergy in Tipperary were doing all in their power to stop boycotting drew a withering retort from Fr David Humphries: 'As one of the Catholic clergy of Tipperary, I protest against this libel on me. I am doing nothing to stop boycotting. I should be very much ashamed of myself if I were'.[24]

The rescript boomeranged upon the Vatican. It caused great indignation amongst the Irish people and severely shook the confidence of the laity in the Holy See.

> 'Irish feeling has been dreadfully wounded', Archbishop Croke wrote to Tobias Kirby on 6 June 1888, 'and the stability of Irish faith severely tested. The Pope is cursed in every mood and tense from Donegal to Baltimore; and wherever his picture was found in private houses it has either been displaced simply, or torn to bits. The Archbishop of San Francisco tells me it is worse still in the United States'.[25]

The reaction of Irish Catholics to the decree and the resultant debate on the question of church-state relations was an impressive display of political maturity and a firm indication that, in an independent Ireland, Home Rule would certainly not mean Rome Rule.

The major portion of Parnell's address at the Eighty Club on 8 May was given over to the Plan of Campaign and his attitude towards it. He claimed that when the Plan was started in October 1886 he was dangerously ill and so unable to express any opinion on it. 'I was so ill that I was unable to put pen to paper or even to read a paper and I knew nothing about the Plan of Campaign until weeks after it had started'. From the agrarian standpoint, he acknowledged, the Plan had been most successful. 'It has prevented enormous suffering in Ireland. It has, in fact, pacified the country'. But he was opposed to it on political grounds. From the beginning, he feared that certain elements in the Plan would have an adverse effect on English public opinion and on the political situation generally.

> 'Therefore', he reasoned, 'I should have advised the Irish tenants to suffer a little while, to bear their evictions in the firm confidence

> that although their troubles might last for a year or two, yet in a short time they might be put an end to forever by the concession of self-government to their country rather than that by embarking in a movement of that kind they should give any power to the enemy to cover over their own iniquity and to raise false issues on questions which had really no bearing on the matter at all'.

If Parnell considered the Plan of Campaign to be a political albatross, why did he not try to suppress it? When he had recovered from his illness at the beginning of 1887, he explained, it was too late to take action. The government had already prosecuted Dillon and O'Brien and, under these circumstances, he was not going to tell them that their agitation had been a mistake. Also, he considered that the Plan had, by then, already worked its mischief. He had, however, stipulated to O'Brien that neither the League nor the parliamentary party should be identified with the Plan, that it should be restricted to those estates where it was already in operation and that there should be moderation in speech and action. He felt that these conditions had been substantially adhered to; with the exception of Dillon's recent activities on Lord Massereene's County Louth estate, the Plan of Campaign had not been advocated on a single property. Any tenants who had adopted the Plan in the intervening period had done so spontaneously and of their own accord. He concluded:

> I believe that in all probability in a short while that the Plan of Campaign in the natural course would have been gradually succeeded by a method of agrarian organisation which I have for some time been engaged in maturing and which would have been, I think, free from those defects, those political defects, which were incidental to the Plan of Campaign and which would have absolutely corresponded in every respect with the system of organisation known as trades unionism in this country. But we shall now have to wait.[26]

Parnell's stipulations concerning the Plan of Campaign were not laid down at the beginning of 1887 but immediately after, and consequent upon, Dillon's violent fulmination against landlordism and the Irish land system at Castlerea on 5 December 1886.[27] Opponents of the Plan who considered it to be both illegal and immoral had their fear and loathing of it much strengthened by Dillon's diatribe. On 7 December, the Marquis of Hartington, addressing the Liberal Unionist conference — derisively dismissed by the *Times'* Irish correspondent as 'a handful of piebald whigs' — said that the Plan of

Campaign 'must lead to the complete and fundamental subversion of every principle upon which hitherto social order has been based'.[28] On the following day, at the City Conservative Club, London, the Prime Minister called the Plan 'this atrocious doctrine', and described it as a system of 'fraud', 'swindling' and 'organised embezzlement'.[29] The landlord Dublin *Daily Express* designated it 'the plan of plunder' while the English Catholic weekly, the *Tablet,* classed it as anarchical and revolutionary.[30]

From its inception, the Plan of Campaign had created difficulties for the Liberal leadership. Privately, Gladstone and Morley were opposed to it, publicly, they debated the underlying causes rather than the agitation itself and blamed the Plan's existence on Tory misgovernment.[31] Dillon's outburst at Castlerea caused considerable disquiet in the Liberal ranks. Gladstone considered it 'very bad'.[32] The influential *Daily News* editorialised on 6 December:

> We by no means approve of Mr Dillon's policy. His Plan of Campaign seems to us to be vitiated with dishonesty... The language used by Mr Dillon at Castlerea... will give comfort to his enemies and distress to his friends. Such speeches are the material with which the opponents of all concessions to Irish claims work on the fears of timid people in England and Scotland. They are wrung forth by a sense of oppression and injustice; but they are used to justify the oppression and perpetuate the injustice.

On the following day, 7 December, John Morley informed Parnell that the Plan of Campaign was having a very adverse effect on English public opinion. Parnell explained that he had been ill and had taken no part in the agitation. 'He was anxious to have it fully understood that the fixed point in his tactics is to maintain the alliance with the English Liberals'. Parnell agreed to send to Ireland at once for one of his own lieutenants and to press for an immediate end to violent speeches.[33]

William O'Brien's colourful description of his fog-shrouded meeting with Parnell behind Greenwich Observatory, sometime between 7 and 12 December 1886, is highly suspect. His account of the agreement hammered out with Parnell bears a striking similarity, in tone and phraseology, as well as detail, to aspects of the latter's address at the Eighty Club in May 1888. O'Brien's reconstruction of the Greenwich meeting appears to be based on the newspaper reports of Parnell's speech. John Morley's account, which he wrote after meeting Parnell on 12 December, is more convincing: 'The lieutenant had come over and had defended the Plan of Campaign. Mr Parnell persevered in his dissent and disapproval, and they parted with the understanding that the meetings should be dropped and the movement calmed as much as could be'. However, it contains one vital flaw; it is inconceivable that William O'Brien would have consented to abandon public meetings in Ireland. The most likely outcome of the Greenwich meeting is that O'Brien managed to reassure Parnell by means of some vague promise to moderate the agitation in Ireland. In reality, Parnell's attempted interference was ignored and the Plan of Campaign was promoted with no less vigour and commitment than heretofore.[34]

A week of so after the Greenwich meeting, a press release was issued on Parnell's behalf, claiming that he had been ill for some time and would not express an opinion on the Plan of Campaign until he had travelled to Ireland to discuss its operation with those responsible for its inception. 'Mr Parnell was not aware that the Plan of Campaign had been devised, or was going to be proposed, until he saw it in the newspapers'.[35] The press was openly sceptical. One liberal newspaper described the statement as 'nonsense... make-believe that could not deceive a Kerry cow'.[36] Wilfrid Blunt rejected the idea of Parnell's illness and suggested instead that he had spent the past three months 'living in idleness with Mrs O'Shea while the rest have been at work'.[37] Dr Lyons assures us that Parnell's illness was 'both genuine and serious' and that he was telling the literal truth when he claimed that he had not been consulted specifically about the Plan of Campaign. However, Parnell was well aware that a land agitation was both imminent and inevitable in the autumn of 1886, even going so far as to canvass funds for the Irish tenants from the president of the Irish National League of America.[38] However the objectives of the new agrarian agitation, enunciated by Dillon at Castlerea on 5 December 1886, were not alone anathema to Parnell's conservative nature but, more significantly, threatened the very basis of the Liberal-Nationalist alliance, the lynchpin of Parnell's political strategy. Thenceforth, his attitude to the Plan of Campaign was one of undisguised antipathy.

Parnell's address at the Eighty Club was the most comprehensive public expression of his attitude to the Plan of Campaign. However, the speech contained many inaccuracies, of which his anachronous Greenwich meeting with William O'Brien was the most glaring. The Massereene tenants had adopted the Plan of Campaign as early as February 1887, while even the most cursory reading of the nationalist press would have shown him that tenants on numerous estates resorted to the Plan after mid-December 1886, often at the exhortation of O'Brien, Dillon and other exponents of the agitation. The assertion that he was maturing a method of agrarian organisation to succeed the Plan, while incontrovertible, was uncorroborated and, in the circumstances, must remain questionable. Finally, Parnell's speech was deeply flawed because its thesis was fallacious. The implication that the problems and hardships of the tenant farmers would have been remedied 'in a year or two' by a Home Rule parliament was sophistical and mischievous. The political events of the summer of 1886 had clearly indicated that there was not the remotest possibility of Home Rule being granted in the foreseeable future.

Parnell's Eighty Club address was a cynical attempt to reassert his authority over a movement which two years of seclusion, political inactivity and, particularly, his failure to lead the agrarian struggle, had done much to undermine. It was an effort to wrest the initiative from Dillon and O'Brien and to indicate that the political struggle took precedence over the agrarian one.[39]

The reaction of nationalist Ireland was one of total bemusement. The *Freeman's Journal* commented, flatulently, on 9 May: 'A franker statement than Mr Parnell has made about the Plan of Campaign it would be impossible to conceive'. In an early attempt at 'disinformation' — 'Mr Parnell referred briefly to the Vatican decree and, incidentally, to the Plan of Campaign, acknowledging its great efficacy in preserving the Irish peasantry from extermination at a moment when the government had rejected all proposals for their relief'—*United Ireland,* on 12 May, described Parnell's speech as 'immensely damaging to the government' while, contrarily, the Irish Under Secretary characterised the statement as 'another blow at the Plan of Campaign'.[40] On the afternoon of 9 May, William O'Brien dashed off a characteristically impetuous and, at times, incomprehensible, article for *United Ireland,* protesting at 'Mr Parnell's reasoning' but bowing to his 'authoratitive judgement'. This article was suppressed by Harrington and Dillon at the proof stage, the latter soberly explaining in a covering note that, had it been published, 'it would have utterly ruined our movement and driven me and others out of public life'.[41]

The Plan of Campaign came through the twin traumas of the papal rescript and Parnell's Eighty Club address more or less unscathed. The Irish party leader's public disavowal of the agitation aggravated the already festering relations between himself and his lieutenants, but it had no appreciable effect on the agitation itself. However, it had been made forcefully clear to Dillon and O'Brien that they could expect neither support nor sympathy from Parnell should they get into difficulties.

The latter half of 1888 was a period of considerable crisis for the Plan of Campaign. The exigency and uncertainty largely arose from the Plan's financial vulnerability, rather than from any external political pressure. The evicted tenants made the greatest demands on the Plan's hoplessly inadequate resources. 'The wounded soldiers of the land war' were housed and given a monthly grant. Frequently, temporary accommodation had to be erected and paid for. Monthly doles and building costs, together with legal, travelling and hotel expenses, proved to be a relentless and ever-increasing financial drain. As early as December 1887, Dillon was worried lest they should be 'stranded for money' while, in the following month, Tim Harrington predicted imminent financial ruin.[42]

Dillon and O'Brien were deeply worried by the continuing high level of evictions on Plan of Campaign estates throughout 1888. Between February and September, many families were ejected from the Ponsonby, Clanricarde, Lewis, O'Grady, Massereene, Tottenham, Cormack, Byrne, Vandeleur and Leader properties.[43] 'What is to be done in this terrible crisis that is upon us?' Canon Keller besought O'Brien on 12 September 1888. 'The landlords are coming down with unwonted violence and probably we are only in the beginning of it'.[44] Colonel Vandeleur, of Kilrush, County Clare, proved to be particularly intractable. The Plan had been adopted on his estate in mid-December 1886. Early in 1888, an attempt was made to settle the dispute but the terms offered by the agent were rejected by the local clergy. In the middle of February, Colonel Turner reported that the tenants were 'really anxious to settle'. A very large amount of arrears had accumulated through mismanagement and other causes. There was no possibility of the tenants discharging their liabilities and Turner feared the consequences of large scale evictions. 'If he does not settle', Turner concluded, 'he will get no money and no rents in the future and the estate must in time be practically lost and valueless'. The Under Secretary was even more direct. He informed Arthur Balfour that Vandeleur was not 'fit to manage his affairs'.[45] The Chief Secretary requested Colonel Vandeleur on two separate occasions to show some restraint but found him 'stupid, obstinate and selfish'.[46] Colonel Turner reported, at the end of May, that the landlord deserved 'little or no consideration and that there was great excitement and indignation throughout County Clare'.[47] Sir Joseph West Ridgeway feared that the whole estate would have to be cleared before the tenants would give in and he confided in the Chief Secretary that it was not a case of which the government could be proud.[48] Some twenty-five families were evicted during the last two weeks of July before the proceedings were postponed indefinitely.[49]

A fortnight later, the battering-ram met its match on the estate of James E. Byrne, Coolroe, County Wexford. 'Somer's Fort' was elaborately barricaded, both inside and out:

> Huge earthworks were thrown up all round the house and outside the earthwork the trenches were dug to a depth of four feet. The upper windows were barricaded with gates and bars, held by chains. Large gates were chained up against the roof, from the inside, as a protection should the house be stripped. Holes were broken through the walls to discharge showers of water on the attacking forces...The laneway approaching Somer's house was blockaded by trees felled during the night so as to obstruct the passage of the battering-ram.

Rotten eggs, boiling water and showers of stones were hurled at bailiffs and police alike. Again and again they were driven back, 'cut, bleeding and beaten'. Baton and bayonet charges were equally unsuccessful and and, finally, the inspector in charge ordered his men to open fire. The local clergy then intervened and, at their instigation, Somers and his friends surrendered. Twelve people were arrested. Willie Redmond MP was subsequently sentenced to three months imprisonment for incitement. The locality remained tense and, on the day after the siege of 'Somer's Fort', Mr Byrne was struck in the face with a large rotten goose egg which, the correspondent of the *Freeman's Journal* gleefully noted, 'besmeared him from head to foot'. The landlord later deposed that, for months after, the mildest salutation he received from his tenants was, 'may you awake in hell in the morning'.[50]

Irish evictions were widely and often indignantly reported in the nationalist and liberal press. Balfour admitted to his cabinet colleagues in November 1888 that the public outcry produced by 'harsh evictions' was 'one of the most formidable difficulties of the government'.[51] But while evictions generated sympathy for the tenants and exploitable political publicity, each hapless victim was an additional burden on the Plan of Campaign's rapidly diminishing resources. An additional problem in mid-1888 was that the leaders of the agitation faced lengthy and simultaneous prison sentences.

John Dillon had been sentenced to two concurrent terms of six months imprisonment at Mel, County Louth, on 9 May 1888, a penalty he thoroughly deserved, opined the Chief Secretary: 'I am glad Dillon got six months. If any man deserved it he is that man'.[52] Dillon's appeal was heard before Judge Kisbey at Drogheda Quarter

SUPPLEMENT GIVEN AWAY WITH THE **WEEKLY FREEMAN** 19TH MAY, 1888.

THE BLACK WATCH.

Brutal Balfour.—"Ha! Ha!! I have got him under lock and key for six months, and what with a delicate constitution and limited doses of skilly, who knows but he may —————————— I mean, he may become a 'Thing of the past,'—like the National League."

Sessions on 20 June. The Judge said that he 'entertained no doubt that what was known as the Plan of Campaign constituted a criminal conspiracy punishable by law and that persons who took part in it were guilty of taking part in a criminal conspiracy'. He upheld the original decision and ordered that Dillon be lodged immediately in Dundalk gaol. An attempt to have the decision reversed in the Court of Exchequer on a writ of habeas-corpus was unsuccessful.[53]

William O'Brien MP came up for trial at Loughrea on 26 April 1888, charged with attending a proclaimed meeting. His counsel determined to call into the witness box every individual who had attended the outlawed meeting to attest on oath that he had done so as a member of the National League. This elaborate and rather tedious farce was brought to an abrupt end when the depositions and other legal documents connected with the case were stolen from the courthouse on the evening of 2 May. The case had to be commenced *de novo* and, on 4 May, O'Brien was sentenced to three months imprisonment.[54]

On 1 July 1888, O'Brien wrote to Dillon in Dundalk gaol that, if his appeal were successful on the following day, he would go either to England or to America to raise funds. He would certainly embark for the United States if Parnell would allow Tim Harrington to accompany him. This was unlikely and it would be impossible for him to face the financial work alone. Relations between himself and Parnell were still 'strained to the point of frigidity'. Tim Healy could be relied on to do them an occasional good turn but Harrington was showing very little interest 'in the work'. 'The more I think of it', O'Brien concluded soberly, 'the less I like the prospect of our being in together, so I shall do nothing foolish'.[55]

As O'Brien must have confidently anticipated, he and Dillon did avoid being incarcerated simultaneously. Under the Crimes Act, the accused was entitled to have produced on appeal the whole of the depositions in the lower court. But they had been stolen and the prosecution could not explain either their disappearance or their whereabouts. The result of this irregularity in form was that the Court of Exchequer sent the case back to be reheard by the magistrates. No further proceedings were taken.[56]

Over the next fortnight, the relationship between O'Brien and Parnell improved considerably. The Irish leader donated at least £2,000 of the money he received from Cecil Rhodes for the relief of the evicted tenants.[57] But this was merely a temporary, if welcome, respite. According to O'Brien, the expenses were 'swelling frightfully'. The Chief Secretary had 'plainly declared war, root and branch, on every campaign estate and undoubtedly, if our grant list goes on swelling, there will be a collapse unless you or I get off on a begging mission'. For a variety of reasons, O'Brien was reluctant to embark on a fund-raising tour and he suggested that the best course would be for Dillon to journey to Australia and America when he was released, 'in a month or so'.[58]

Dillon's health, always suspect, had deteriorated in prison, giving rise to many furious gibes by the Chief Secretary on the connection between 'diseased lungs and Irish patriotism'.[59] Fearing, in Salisbury's brutal phrase, that Dillon would prove 'more formidable dead than alive', the government released him in mid-September, having served slightly less than half of his six months sentence.[60] Dillon's reprieve lifted some of the awesome pressure from O'Brien–'The anxiety is sometimes too much for my pair of shoulders', he had confided to Dillon on 16 July—but it did nothing to ease the Plan's financial difficulties.

On 28 October 1888, Dillon recorded that he had just received a summary of Plan expenditure from Laurence Ginnell. A sum of £1,261 was being spent on monthly grants to support 312 tenants and 92 labourers. 'That means an expenditure of £20,000 a year — all items included', he noted.[61] A week later, O'Brien and himself spent two gloomy hours discussing the situation. They estimated that they would need at least £25,000 to maintain the agitation for another year and, unless Parnell came to their aid, they would be unable to continue. 'The situation is doubtless one of tremendous difficulty', he concluded despondently.[62] Their position became even more critical during the winter. The new year was marked by the resumption of evictions on the Olphert estate in Donegal. The owners of the Tottenham, Stewart and Kenmare properties followed suit before the end of January. In mid-January 1889, a desperate John Dillon appealed to Parnell for £10,000, half immediately and the remainder before 1 March. If their position could be secured until then, his mission to Australia and a similar fund-raising trip by O'Brien to America should raise enough to bring the Plan of Campaign to a successful conclusion.[63] On 26 January, Parnell churlishly agreed 'to scrape together £10,000 for you'. He observed, more benignly: 'I was delighted to see how cleverly O'Brien escaped and I hope he will lose no time in visiting America and collecting some money for your needs, avoiding any further risks before he starts'.[64]

Despite the Plan's acute financial problems, O'Brien had stressed, in his letters to the imprisoned Dillon during the summer of 1888, that the spirit of the people was 'simply glorious'.[65] The indomitable

O'Brien was determined 'to give Balfour another hot winter' and suggested that the Irish members should organise a series of meetings in their own divisions, 'so as to have the heather thoroughly on fire in October and to show Balfour that his work is only beginning'.[66] On 25 August, he informed the Killarney MP and hotelier, J. D. Sheehan, that the meetings were scheduled to begin towards the end of September and urged him to arrange popular demonstrations 'from parish to parish' throughout his constituency.[67]

On six successive Sundays between 23 September and 28 October 1888, major meetings were held in at least twenty-five centres and, with the exception of Dillon who was recovering from his recent prison ordeal, were addressed by virtually every MP who had been associated with the Plan of Campaign from the beginning.[68] O'Brien made the following observation at Ballyneale, County Tipperary, on 30 September: 'When landgrabbing broke out in a district every honest man should feel as if an epidemic of yellow fever had broken out... He ought not to rest a single day until the last trace of it is stamped out and until the very air is purified of the contamination'. He was subsequently charged with conspiracy to prevent the hiring of farms.[69]

At Mount Horgan, County Kerry, on 28 October 1888, Lord Kenmare's tenants adopted the Plan of Campaign as an antidote to 'merciless rackrents and intolerable landlord oppression'. They were seeking a reduction of twenty per cent on judicial rents and forty per cent on non-judicial ones and pledged themselves to adhere to the Plan until their 'reasonable demands' were met. O'Brien said that the Kenmare estate was an example 'of all the worst crimes and the cruellest oppression of Irish landlordism'.[70] On 23 November, Colonel Turner reported to the Inspector General of the RIC that, since O'Brien—'that most pestilential agitator'—had written to Sheehan, the whole county of Kerry was seriously disturbed. 'Moonlighting, intimidation, the Plan of Campaign and murder have been the result of his interference... The meetings held by Sheehan and others in pursuance of O'Brien's instructions have led entirely to the present state of things... I have grave apprehensions as to troubles to come'.[71] O'Brien addressed another Plan of Campaign meeting on the Kenmare estate, on 20 December 1888, and, as a result, was charged with taking part in a criminal conspiracy.[72]

On 25 January 1889, at Carrick-on-Suir, O'Brien, in his absence, was sentenced to four months imprisonment on the first charge extant against him. On the previous day, after the Crown had concluded its case, Tim Healy applied that either Lord Salisbury or

Arthur Balfour be summoned to appear before the court. Thwarted by the magistrate in his intention, Healy said he would call the whole county of Tipperary to give evidence. He then referred to George Bolton, the Crown solicitor, and there was an immediate 'outburst of cheering and groaning'. The magistrate ordered that the public gallery be cleared. O'Brien and Healy said that if this were done they would leave the court, 'and thereupon they did leave, Mr O'Brien suddenly dashing through the police and darting up a by-street amid the confusion'. In disguise, he escaped to Wexford, crossed to Wales and made his way circuitously to Manchester, where he was scheduled to speak on 29 January. He successfully evaded 'the entire police force of Manchester' and, when he made his dramatic appearance before the auditorium of the Hulme Town Hall, as he triumphantly described it, 'the house arose, danced, sang, wept, embraced each other and for better than a quarter of an hour went raving mad with joy'. O'Brien was allowed to deliver his address and, on the following day, 30 January, he was returned to Ireland and lodged in Clonmel gaol. The police were convinced that the Carrick-on-Suir escapade was a pre-arranged and carefully-orchestrated plot to add to O'Brien's notoriety. The author of the report commented tartly that this 'regular little drama... like most other Irish political episodes' was enacted for effect on English public opinion.[73]

O'Brien's appearance at Tralee Petty Sessions on 18 February 1889 was marked by the suspension of his counsel, Tim Healy, for improper conduct in calling Colonel Turner 'a sneak'. The defendant was sentenced to six months imprisonment, which a unionist weekly described as 'tidings of comfort and joy to all in Ireland who want to mind their own business'.[74] The sentence was reduced on appeal to terminate coincidentally with the one he was then undergoing.[75]

In Clonmel gaol, O'Brien had renewed his fight for the right of political prisoners to wear civilian clothes. A bitter conflict ensued with the prison authorities which caused Dillon to postpone his departure for Australia.[76] O'Brien managed to reassure him and urged him to leave immediately.

> *It is all right now'*, he wrote to Dillon at the end of his first week in prison, 'I believe I have practically won. That being so I beg of you to go at once. Want of money alone could beat us. I will face America when I come out. It was an ugly ordeal but, thank God, the worst is over and the flag is flying'.[77]

On 21 February 1889, O'Brien was transferred to Galway gaol and under its more sympathetic administration embarked on his first novel, *When We Were Boys*. A fortnight later, Dillon set off on his thirteen-month begging mission to the Antipodes. In the meantime, a rather innocuous report in a Cork unionist newspaper of a dinner held in the Imperial Hotel to honour A. H. Smith-Barry and R. U. Penrose Fitzgerald presaged the climactic struggle of the Plan of Campaign agitation.[78]

7 *Deus Ex Machina*

If I were an Irish landlord I would beg my bread before I gave in to the Plan of Campaign... and I would beg my bread rather than let any other landlord give in for want of support.
Arthur James Balfour, in the House of Commons, 30 January 1891.

I am well aware of the pressure to which every landlord is subject in carrying on the fight against the Plan of Campaign. But I would most earnestly point out that it is absolutely vital to the interests of the landlords that they should make common cause to aid each other... It is by combination that the tenants have been successful; and, in my opinion, it is only by combination that they can adequately be met.
Balfour to Lord Courtown, 2 May 1888 (BL, Balfour Papers, Add. Ms. 49826).

(i)

The reaction of the landlords to the Plan of Campaign earned for them the obloquy and disapprobation of successive Chief Secretaries at Dublin Castle. Sir Michael Hicks Beach had urged property owners to take bankruptcy proceedings against defaulting tenants but, he complained to Lord Salisbury in mid-December 1886, 'Irish landlords... prefer to lie on their backs and look to the government'.[1] Arthur Balfour decried the lack of spirit among the landlords and described them as 'contemptible... maddening clients... fools... They always cry out before they are hurt when the government is concerned but when the National League is concerned they fold their hands and do nothing'.[2]

A number of protective organisations, some of which dated back to the days of the Land League, provided some relief for the beleaguered landlord garrison. The Property Defence Association was founded in December 1880 'to uphold the rights of property against organised combinations to defraud and to maintain freedom of contract and liberty of action'. All individuals, regardless of creed or political opinion, who were interested in the maintenance of law and order and anxious to halt the advances of communism were invited to join. The organisation supplied men to service writs and other legal processes; sent agents to bid at sheriffs' auctions at which tenants' stock and interests in their farms, seized for non-payment of rent, were put up for sale; tenanted evicted farms; arranged for the disposal of the stock and produce of boycotted individuals and provided labourers for them; and provisioned persons whom local traders refused to supply. In 1886, Lord Courtown, the chairman of the association, referred to the Plan of Campaign as 'the newest device of the League for defrauding landlords; and insisted that continued attacks on property were inevitable. 'Those who have subsisted on public plunder are not likely to give up their profitable occupation', he opined. In the following year, he reported that the Plan had been widely adopted, 'with consequences disastrous to the interests of both landlords and tenants'.[3]

The Land Corporation of Ireland Limited was founded by the remarkable Arthur McMurrough Kavanagh in July 1882 and was incorporated on 17 March 1883 with a share capital of £720,000, divided into 140,000 ordinary shares at £5 each and 4,000 deferred shares of similar value. The ordinary shareholders were guaranteed an annual minimum dividend of 5 per cent and to secure it the Land Corporation of Ireland Guarantee Company Limited was formed and registered with a capital of £350,000. According to Kavanagh, the primary object of the Land Corporation was to counteract the Land League and later the National League and to complement existing landlord organisations, such as the Property Defence Association and the Emergency Committee. 'These two associations', he said, 'enabled the landlords to assert their rights. Our object was to take up their farms after this had been done and work them'. A full range of activities associated with a land and financial company were undertaken including land agency and estate management, the commercial working of evicted farms and the provision of loans to proprietors and tenants of Irish land.[4]

The Land Corporation played a singular role in combating the Plan of Campaign on the estate of The O'Grady at Kilballyowen, County Limerick. In February 1887, T. J. O'Shaughnessy, 'oil, color, seed, timber, and general hardware merchant' of Bruff, advised Madam Elizabeth O'Grady that he could no longer supply her household 'under the present excited state of public feeling'. He recommended that The O'Grady try to effect a settlement with his tenants, 'even at a loss'. O'Shaughnessy also wrote to The O'Grady's gamekeeper and to James O'Mahony, 'at The O'Grady's', requesting that neither they nor their families purchase any goods in his shop until the dispute was settled. 'I needn't tell you it is most painful to me to be obliged to have recourse to this proceeding, but such is the public feeling'. Another Bruff shopkeeper advised Madam O'Grady to get her 'stuffs' elsewhere, lest any trouble 'obstruct this town'. M. Carroll, who had 'enemies in this village as well as anybody else', also refused to supply The O'Grady. 'I am trewly sorry for to have to take the above step but I neede̱nt tell you that public opinion cannot be ridden over'. Madam O'Grady's dressmaker, J. O'Sullivan, returned her 'habit undone owing to present unpleasant circumstances', but fervently hoped that 'O'Grady will see his way to remove those impediments which prevent me doing everything in my power to oblige you in doing anything you should require me to do for you'. By August 1887, The O'Grady's entire household was boycotted, his servants were prevented from attending church, his labourers had been driven away and his agent had been forced to resign.[5]

Arthur McMurrough Kavanagh informed his niece, Madam Elizabeth O'Grady, that the Land Corporation regarded theirs as 'a special test case' and agreed to help them 'as a matter of public policy'. They appointed Arthur White, a Limerick land agent, to act on their behalf and gave him £500, interest free, to stock the estate. The Land Corporation undertook all the expenses of farming the property and agreed to pay seventy-five per cent of the net profit to The O'Grady as rent. In return, he gave a written undertaking not to settle with his tenants other than on terms approved of by the Land Corporation. In August 1887, a new organisation, the Anti-Plan of Campaign Association,[6] undertook to assist The O'Grady in every possible way on condition that he held out against his tenants. It was felt that a surrender to the Plan of Campaign on this estate, where its adoption was considered to be particularly unjustified,[7] would have a very deleterious effect on the country generally. It was on the basis of this understanding that The O'Grady agreed to fight the Plan, as his difficult financial situation precluded him from doing so on his own account. From then, White recorded in 1896, all of his actions were governed by his agreement with the association.[8]

The Cultivation of Derelict Land (Ireland) Trust, which was set up at the end of April 1889, was inspired by the highly individualistic attempts of Lord Massereene to counter the effects of the Plan of Campaign on his Louth and Meath estates by planting thirty-two evicted holdings with Protestant tenant farmers, mainly from Ulster. Advertisements in the northern newspapers in December 1888, stipulating that 'none but Protestants need apply', offered 'special advantages' to suitable applicants. Rent was waived for the first six months and the gross rental on the thirty-two farms was reduced from £872.50 to £820.75; the dwelling houses and the farm outbuildings were repaired and improved; loans amounting to £650 were secured for five tenants.[9]

Lord Massereene justified his actions on the grounds that not a single Protestant tenant on his estates had joined the Plan of Campaign. He denied that his motives were sectarian. He was simply determined 'to obtain tenants who have never been and are not ever likely to be connected with such organisations as the National League and the Plan of Campaign'.[10] However, the scheme was widely denounced. George Shaw Lefevre warned in the *Times* that it would 'revive and confirm religious animosities'.[11] The parish priest of Monasterboice described the new tenants as 'the riff-raff of Ulster' and 'Orange tramps'.[12] The clergy of the deanery of Drogheda expressed 'extreme concern and regret' at Lord Massereene's policy. They maintained that it was insulting to the Catholic religion and likely to cause sectarian strife which would disturb 'the peace and tranquility of a hitherto crimeless district'.[13] The Under Secretary, Sir Joseph West Ridgeway, believed the 'colonization' scheme to be a good thing *'pour encourager les autres'* but was reluctant to see it recklessly extended, lest it end in a religious war. It would certainly cause bitter sectarian feeling and enrage the priests, he concluded.[14]

Clearly influenced by Lord Massereene's scheme, T. W. Russell, Unionist MP for South Tyrone, wrote an article for the *Times* on the working of the Plan of Campaign on the estate of George F. Brooke, Coolgreany, County Wexford.[15] Consequently, several individuals offered to subscribe to an organisation to aid suitable tenants to take vacant farms on Plan of Campaign estates.[16] The result was the formation of The Cultivation of Derelict Land (Ireland) Trust with Walter Gyles, secretary of the Land Corporation, as chief director.[17]

It was intended to plant the Leader, Brooke, O'Grady and Lansdowne estates. The seventy farms on the Brooke property from which tenants had been evicted were distributed among twenty-one new tenants, all but two of whom were Protestants. Seven came from Wexford, four from Cavan, two each from Kildare, Wicklow and Dublin, while the remainder came from Antrim, Carlow, Queen's County and Aberdeenshire.[18] With the exception of the two Catholics, who received no reductions, the rents were lowered considerably, in one case from £116 to £100, in another from £439 to £370, while a third was reduced by £16 to £74.[19] The tenants were given a loan of £3, at an annual interest rate of three per cent, for each acre they farmed. The capital sum was to be repaid in ten triennial instalments. They were given thirty-one year leases which they deposited as security with the Land Trust.[20]

The Brooke plantation was not a success. A land valuer named Richard M. Barrington, who inspected seventeen holdings numbering 2,400 statute acres in late 1892, found that most of the land was 'mismanaged and deteriorating in value'.[21] In October 1893, the local RIC inspector reported: 'The planters have as yet paid neither principal nor interest. They all received letters towards the end of September last demanding payment up to date under threat of immediate proceedings'. A deputation met the company's secretary to explain 'the impossibility of paying the money at present'.[22] The scheme on the Massereene estate was much more successful but, even here, the lands were not farmed to their full potential. One of the reasons given was the considerable difficulty under which the planters operated. It was said that they needed 'the constant protection of a considerable force of police'.[23]

While it was conceded that the various landlord organisations did useful work in defending property rights, it was widely acknowledged that the Plan of Campaign could only be effectively countered by the Irish landlords combining for mutual support in a national organisation. As early as December 1885, when the Cork Defence Union was formed—primarily to assist boycotted individuals and all who were opposed to the policies of the National League[24]—A. H. Smith-Barry had called on the landlords to emulate the nationalist example: 'They should organise and stand shoulder to shoulder as the Land League did and they should support each other when in trouble.[25]

The unionist press quickly came to the same conclusion. The Plan of Campaign was anarchical and a threat to the foundations of civilized society, postulated the *Times* on 20 January 1887. It could only be overcome by 'counter combination'. On 16 November 1886, in an editorial comment on the adoption of the Plan by the Ponsonby tenants, the Dublin *Daily Express* suggested that the Ponsonby estate ought be treated as a test case:

> If the tenantry, urged on or backed by the National League, be resolute on their side, so should be the landlord on his, and as the tenants have an organisation to sustain them so should this landlord, if he have not the necessary means for the struggle, be supported by the united strength of the class.

On 11 December 1886, two days after the *Daily Express* had

repeated its call for property owners to form 'a counter organisation to meet the League', the County of Cork Landowners Association was established at the prompting of A. H. Smith-Barry. Its disclosed aim was 'to combat the Plan of Campaign as proposed in *United Ireland* and for the protection of landed property in the county generally'.[26] From the beginning, the association closely identified itself with the struggle on the Ponsonby estate. On 15 December, a meeting chaired by Smith-Barry resolved to make arrangements 'without delay for dealing with the case of Mr C. W. Talbot Ponsonby'.[27] An appeal for funds was launched in the *Times* by the newly-elected chairman, D. R. P. Sarsfield:

> I appeal to all those who desire to see law maintained and their country saved from absolute anarchy to assist with subscriptions to defray the necessary expenses in the contest which has been forced upon us by the enemies of England and of Ireland. The expenses will be considerable but we feel that in contesting this case we are fighting the battle of all owners of property of any sort in Great Britain and Ireland.[28]

The sauce that coated the nationalist goose was deemed an unacceptable condiment for the landlord gander. The appeal for funds drew a passionate rejoinder from an irate William O'Brien. At a meeting of the Ponsonby tenants at Inchiquin on 6 March 1887, he stormed:

> We have come here today to throttle one of the most atrocious conspiracies that was ever formed against the homes and the happiness of a body of industrious Irish tenants, a conspiracy to exterminate our people and insult our priests...It is not Mr Ponsonby we are fighting in this contest; it is the gang of landlord conspirators in Cork who know nothing of this estate...It will never be forgotten for those men until we have them howling for mercy at our feet. This is as bad and base a crime as ever was laid to the charge of landlordism in all its bloodstained and miserable history...bad instincts of landlordism...incorrigible hatred of the Irish people...a smouldering volcano...battle and war to the knife...tormentors of discord and injustice...the Ides of March are approaching...If one hair of Fr Keller's head is injured or if one of the houses on the Ponsonby estate is unroofed we will throttle them at their own rent offices...and kindle a blaze about their ears.

O'Brien concluded this uncharacteristically vitriolic tirade by urging the nationalists of County Cork

> to mark every man who allies himself with this band of priest hunters in Cork, who attends their meetings, who subscribes to their funds, and mark him well... Don't harm a bone in their miserable carcases... Draw a ring of fire around every man of those, a circle of excommunication. Boycott them. Stop their hunting. Don't deal in the same shop with them. Make them feel every hour of the day and every moment of their lives that they themselves have proclaimed war, a war of extermination on the people and that they themselves bear the guilt and the brand of the exterminator on their brows.[29]

O'Brien's fulmination convinced Smith-Barry that law-abiding citizens had no option but to take up his challenge and 'fight it out'.[30] A fortnight later, in the *Union*, a recently established weekly primarily designed to counteract *United Ireland* propaganda, he denounced the Plan of Campaign as a 'communistic agitation', intent on the spoliation of all types of property. Its primary aim was to drive the landlords— a considerable obstacle to the repeal of the union'—out of Ireland. It was essential to counter such an agitation by 'mutual co-operation' among the propertied classes. 'If those in Ireland who love order and peace wish to see them restored they must pull together and act together and work together'. Smith-Barry flailed the prevailing apathy of the unionist population. 'There is a prevalent idea that all will come right; that someone from outside will, somehow, see us through it all and that it is unnecessary—nay imprudent—to help ourselves'. He believed they were engaged in a vital struggle and defeat would mean the loss of their homes, their properties and their liberty. He urged 'the loyalists of the three southern provinces' to join together in a common cause. By so doing, he concluded, they would not alone win 'the respect and sympathy of the rest of our fellow subjects in all quarters of the globe' but would secure 'the triumph of Liberty and of Law against Anarchy and Terror' and would see once more 'a happy and peaceful Ireland'.[31]

Smith-Barry's powerful plea fell on deaf ears. The Irish landlords, as Balfour informed his uncle, continued to be 'the despair of their friends'.[32] When he heard of the abject surrender of Sir Henry Burke to the Plan of Campaign in February 1888, the Chief Secretary exploded: 'It is utterly useless to try and help the Irish landlords by trifling grants from the treasury when they show themselves so utterly incapable of the simplest combination for the purposes of self-defence and to permit themselves to be destroyed piecemeal in this fashion'.[33]

Balfour's attitude to the Plan of Campaign closely resembled that of Smith-Barry. He variously described it as 'robbery sanctioned by boycotting' and as 'highway robbery'. Its objective was the destruction of landlordism and the transference of land ownership to the tenants. 'If the Plan of Campaign were not fought', Balfour argued, 'it would in six months not merely diminish but wholly destroy the value of every landlord's property from one end of Ireland to the other'. The agitation was part of the political and social revolution that had begun in 1879 and, as such, he claimed, the tenants were mere pawns in the hands of political agitators. They, no less than the landlords, were victims of the Plan of Campaign. The refusal to pay rent had resulted in numerous evictions. The farms, left idle, had deteriorated and many former tenants now depended on charity for their survival. The landlords were deprived of their rents and their families and creditors were penalised accordingly. Balfour concluded that the interests of the community at large dictated that the Plan of Campaign should be met with 'unyielding resistance'.[34]

(ii)

By the beginning of 1889, there were definite grounds for government optimism. Police reports for December had been most encouraging. Counties Clare and Kerry had shown a 'marked and steady improvement' during the course of the year. Mr Byrne of the western division was uncharacteristically expansive: 'Everywhere I go I see a marked improvement in the disposition and feelings of the people... The law has been thoroughly vindicated... offences against the Crimes Act are becoming less frequent... I look forward to the coming twelve months with every expectation of still further improvement'.[35] In jubilant post-Christmas mood, the Chief Secretary informed his uncle that 'everyone here is in high spirits'.[36] In a lengthy cabinet memorandum, drawn up in February 1889, Balfour claimed that agrarian crime was lower than it had been at any time since 1879. The National League had lost a great deal of its power and was declining in most parts of the country. The attitude of the people towards landlords and police had changed from one of 'sullen defiance to comparative cordiality and goodwill'.[37]

The Plan of Campaign was, by the beginning of 1889, faced with considerable difficulties. There was a crippling shortage of funds, Parnell was hostile, Dillon had embarked for Australia, O'Brien was in gaol and was not due to be released until 22 May 1889. In the meantime, O'Brien's determination to set the heather ablaze for

Balfour in the previous autumn had led to the prosecution of several of the more committed exponents of the Plan. Between 30 January and 7 March 1889, David Sheehy, Joseph R. Cox, Tom Condon, John O'Connor, John Finnucane, J. L. Carew, Dr Tanner and Denis Kilbride were sentenced to prison terms ranging from three to six months on conspiracy charges under the Crimes Act.[38]

However, the agitation was far from moribund. In the early months of 1889, settlements were arrived at on ten Plan of Campaign estates, all of them favourable to the tenants. In at least two instances, the landlords settled on positively penal terms. The Rev. Conyngham Ellis of Abbeyfeale, County Limerick, and W. Fitzgerald of Corofin, County Clare, reduced their rents by forty per cent, cancelled all arrears, reinstated the evicted and paid all legal expenses.[39] The Clanricarde estate continued to embarrass the government. During the winter of 1888–9, Balfour drafted, but subsequently abandoned, legislation to place the property of obstinate landlords, like Clanricarde, under the control of the High Court, for the common good.[40] The evictions on the Olphert estate at the beginning of January, that led indirectly to the death of District Inspector Martin and the prosecution of Fr James McFadden, attracted a great deal of attention.[41] Major Plan of Campaign struggles still continued on the estates of Lord Clanricarde, Lord Massereene, Lord Kenmare, The O'Grady, C. W. T. Ponsonby, Wybrants Olphert, George Brooke, Colonel C. G. Tottenham and Lord Lansdowne. The latter five constituted the so-called 'test estates', properties on which the Chief Secretary was determined that the Plan should not triumph. Of the five, the Ponsonby estate gave the greatest cause for concern.

When C. W. T. Ponsonby, a retired English naval officer, inherited this 10,367 acre property, situated near Youghal, County Cork, in 1868, his new tenants presented him with an address of welcome: 'The name of Ponsonby is traditionally revered in this part of the country, being associated in the recollections and impressions of the people with all that is exalted, honourable and generous'.[42] By the mid-1880s, relations between Ponsonby and his tenants had deteriorated dramatically. In the winter of 1885, a three-week rent strike forced the agent to concede reductions ranging from ten to twenty per cent. Similar abatements were offered in the following year but the tenants demanded that the judicial rents be reduced by twenty-five per cent. When the landlord refused, they adopted the Plan of Campaign on 14 November 1886.[43]

Landlord sympathisers claimed that the property was heavily encumbered and that Ponsonby would be unable to hold out indefinitely against the Plan.[44] By the summer of 1887, arrears of £20,000 had accumulated. In the following November, Balfour informed his uncle that a victory for the Plan on the Ponsonby estate would diminish the value of every Irish property by two years purchase. 'I have done all I could with decency to get this put right but, short of giving money myself, I have little hope of effecting anything'.[45]

In December 1888, J. E. Brunker, a former agent on the estate, attempted to bring about a settlement. He proposed that the property should be sold to the tenants for £127,000. On Christmas Eve, Canon Keller made a counter-offer of £100,000. Early in the new year, District Inspector Ewart reported that there was a 'very general feeling that the matter will now be settled as all sides are very anxious that it should be'. A fortnight later, the asking price was reduced to £110,000. Keller responded by increasing his offer to £106,000.[46] On 26 January 1889, Ewart noted that the difference between the two sides had been further reduced 'and there is a general feeling that an arrangement will be arrived at'. County Inspector Heard reported that a settlement was imminent. 'It would be a great public advantage if the dispute were settled without giving way too much to the tenants', he said.[47]

Balfour considered that the terms offered by the landlord represented a complete capitulation to the Plan of Campaign. He informed his uncle, on 18 January 1889, that he was 'much perplexed' by the situation on the estate:

> Ponsonby is completely "broke" and has lost all nerve. He has no money to pay his charges and none to furnish himself with bread and butter. I must try and get him some by hook or by crook, on condition he fights on. But I am doubtful of success. I am working the thing to the best of my ability. But...I have to keep in the background and the want of funds hampers me seriously. If Ponsonby gives in it will be the worst thing that has happened for some time past.[48]

Within days, the Chief Secretary was maturing a scheme to involve a number of wealthy English and Irish landlords in a syndicate to purchase the Ponsonby estate. By the first week of February, the Dukes of Norfolk, Devonshire and Westminster, Lords Fitzwilliam, Ardilaun and Derby, Walter Morrison and A. H. Smith-Barry had agreed to contribute £10,000 each. The Chief Secretary was hopeful that the Duke of Bedford, Viscount De Vesci, the Earl of Pembroke and Lord Brownlow would do likewise. Lord Revelstoke had pledged himself for £5,000. Balfour impressed on the various members the absolute necessity for secrecy. He was particularly concerned that his own name should 'never appear in connection with private efforts to oppose the Plan'.[49]

On 14 February, Balfour informed Ridgeway that the Ponsonby estate had been purchased by the syndicate and that the affair was 'satisfactorily settled'.[50] On that same evening, the unionists of Cork city and county gave a banquet in the Imperial Hotel in honour of A. H. Smith-Barry and R. U. Penrose Fitzgerald, 'in recognition', as the local unionist newspaper put it, 'of their distinguished services on behalf of the union and particularly the claims of the loyalists of the south of Ireland'. Smith-Barry, in returning thanks, said:

> These are time of great historical crisis in the annals of our country and it is necessary for all of us to do our share as best we may to avert the great danger that threatens our country. This is a duty from which we cannot shrink... The sense of duty will encourage us to continue with the fight we have started in, to continue fighting the cause of our country until that cause is fully and firmly won.

He believed that the government's law and order policies and, in particular, the way they were being implemented by 'that great statesman', Arthur Balfour, were bearing fruit. There was a noticeable improvement all round. Agrarian crime and boycotting were decreasing and the Plan of Campaign was 'not succeeding'. The Vandeleur, O'Grady and Brooke estates were 'safe', W. N. Leader was 'holding out gallantly' at Curass and 'with regard to the Ponsonby estate', he concluded without the slightest trace of irony, 'I am confident that those who instituted the Plan were further off from defeating Mr Ponsonby than they were at the beginning'.[51]

In response to an urgent inquiry from Canon Keller, Brunker telegraphed that Ponsonby had not authorised Smith-Barry's statement. Within a week of the syndicate purchasing the estate, the negotiations between Keller and Brunker were formally broken off by the trustee of the property, a Mr Martineau: 'Having fully considered offer made by Canon Keller and the sum being so far below price named by Mr Ponsonby, trustee cannot advise him to accept'. Brunker, whose motives appear to have been genuinely altruistic, was upset and apologetic. 'I deeply sympathise with you in the disappointment which you must feel on finding that our well-meant efforts have been in vain. Kindly say farwell to the tenants for me and express to them my regret that I could not be of more effective service to them'.[52]

On 21 February, the *Freeman's Journal,* citing a news agency report, announced that the Ponsonby estate had been sold 'to an association of English capitalists who will take possession of the property and devote it to profitable purposes without reference to the Plan of Campaign'. On the following day, District Inspector Ewart confirmed that negotiations between Ponsonby and his tenants were at an end. They had been taken by complete surprise. According to County Inspector Heard, the tenants were so sure of a settlement that many had begun to till their farms. 'If a company has purchased the estate', he concluded sententiously, 'I trust it will be worked vigorously, as it is most demoralising to see farms lying waste'.[53]

Arthur Hugh Smith-Barry and Walter Morrison were chosen to act on behalf of the syndicate. Articles of Agreement, dated 2 April 1889, were signed with Mr Ponsonby by which he agreed to sell his estate for £71,433, subject to mortgages of £22,200 and to the annual charges on the property. The sum of £20,320 was paid on account and a further £25,000 between then and 28 March 1896, leaving a balance of £26,113 due on that day.[54] One of the conditions of purchase was that Mr Ponsonby was to remain the nominal owner and all legal proceedings were to be taken in his name.[55] Smith-Barry informed his constituents at St Neoths in July 1889 that 'an arrangement' had been made with regard to the property but no assignment had taken place and 'Mr Ponsonby therefore remained its legal owner'.[56]

Smith-Barry made no attempt to hide his involvement with the syndicate. Morrison disclosed his membership in a speech to his constituents at Skipton on 21 January 1890.[57] Balfour's seminal connection, long suspected, was never publicly acknowledged or proved. During question-time in the House of Commons on 12 and 13 March 1889, he refused to be drawn on the Ponsonby estate. He admitted to W. J. Lane that he had seen a report of the sale of the property to a London syndicate but that he had 'no official knowledge on the subject'.[58] By implication, however, he left no one in any doubt as to where his sympathies lay. During a motion of adjournment on 1 July 1889, the Chief Secretary informed the assembled MPs that, if he were an Irishman and found individuals combining against him, he would not hesitate to conspire against them.[59]

The charge most frequently levelled at the syndicate was that it had deliberately sabotaged what was seen as the imminent settlement between Ponsonby and his tenants.[60] Smith-Barry responded that such a conclusion was both premature and problematical. In addition to the £4,000 that separated the two sides, there were government charges—quit rents, tithe rents, drainage loans—for which the tenants were responsible. These were estimated at £18,800, which would have pushed up the purchase price to about £129,000.[61] But the question of price was not the fundamental issue. Addressing his constituents at Brampton in May 1889, Smith-Barry made it perfectly clear that the essential purpose of the syndicate was to defeat the Plan of Campaign on the Ponsonby estate and the cost of the undertaking was largely irrelevant:

> The Plan of Campaign must be broken down and defeated. He and his friends had joined a syndicate to defeat the Plan of Campaign on the Ponsonby estate and...they were determined that the conspiracy should be defeated...Law and justice was what every citizen ought to endeavour to maintain and they were determined to do so, cost what it might.[62]

On 3 April 1889, District Inspector Ewart reported that the Ponsonby tenants were apprehensive and that there was 'a strong desire among most of them to settle'.[63] But when they were offered alternative rental and purchase terms two days later, they rejected them out of hand. An indignant Canon Keller claimed that they were less favourable than those offered by the agent at the beginning of the dispute in November 1886; they were inferior to the proposals put forward by Ponsonby in October 1887 and 'they fell immensely below' the terms which Brunker was prepared to accept on the landlord's behalf a few months earlier. Keller insisted that 'they could not have been seriously meant and were simply regarded as a mockery'.[64]

During the summer and autumn of 1889, R. A. Atkins and J. P. Hamilton, respectively the High Sheriff and Recorder of Cork, acted as peacemakers. However, their intervention was unsuccessful due to the insistence of the tenants that the landlord repair all dwelling houses and outbuildings on evicted farms, that each side pays its own legal costs and that any disputed points be submitted to arbitration.

> 'If he yielded on these particulars', Hamilton informed Atkins, 'it would at once be proclaimed as a complete victory by the advisors of the Plan of Campaign. Now this the supporters of Mr Ponsonby will never yield to. I tell you the great question was not a matter of money but of giving triumph to the Plan of Campaign...Anything that could be twisted into yielding to the Plan he will not concede, whatever be the consequences'.

A bemused Canon Keller commented: 'Our conclusion was that they required the tenants to go down on their knees and make a public act of contrition'.[65]

During the next two years, individuals as disparate as Archbishop Croke of Cashel, Bishop O'Dwyer of Limerick, Count Arthur Moore, Sir John Arnott, a Scot named McClelland, a friend of Smith-Barry's from Cheshire, Colonel Eyre, MP for Gainsboro, John O'Connor MP, T. W. Russell MP and Mr Dickson MP, all unsuccessfully intervened in the dispute. Arnott proposed to purchase the estate from the syndicate with the view of subsequent resale to the tenants. He believed that the difference between the two sides could be reduced to about £10,000 and he was prepared to make up the shortfall. However, the syndicate insisted that Park House and demesne be included in the sale and demanded a minimum bid of £150,000. A chastened Sir John did not pursue his altruistic zeal any further.[66]

Compromise was outside the syndicate's terms of reference. Its salaried agent, Horace Townsend, made it perfectly clear that, if the tenants adhered to the Plan of Campaign, the new owners were prepared to clear the estate and farm the evicted lands, using the most modern methods and machinery. It was asserted, and Townsend did not deny it, that he had been given £10,000 to stock the lands and that the members of the syndicate were prepared to forego profits for four years.[67] In either case, whether the tenants capitulated or were evicted, the Plan of Campaign faced inevitable defeat on the Ponsonby estate.

(iii)

Fr David Humphries of Tipperary castigated the *Freeman's Journal* on 7 March 1889, a fortnight after the existence of the syndicate had been publicly exposed by that newspaper, for publishing 'landlord cock-and-bull stories' which were clearly inspired by the landowners to frighten their tenants into submission. 'A London company buying evicted estates in Ireland! Do you believe this trash?', he demanded incredulously. 'Is the *Freeman* to be used as an agency to circulate the ideas of the Irish landlords for the demoralization of the Irish tenants?' Initial disbelief gave way to indignation and protestation but, given the enormity of the landlord undertaking, the nationalist reaction was surprisingly muted. This phenomenon is largely explained by the absence of Dillon and the continuing imprisonment of O'Brien and several of the leading campaigners. It was not until the second half of June 1889, more than four months after the formation of the syndicate, that a discernible nationalist backlash developed.

Between 18 and 22 June, a series of evictions on the Ponsonby estate left some forty tenants, sub-tenants and labourers, together with their families, homeless. There was a great deal of excitement and considerable conflict.[68] The *Cork Examiner,* in a strongly-worded editorial on 24 June, described the evictions as 'this ignoble war on the people' and strongly denounced police brutality:

> On Saturday night, when the forces returned with their prisoners from the seat of war, a disturbance was manufactured in the streets of Youghal and batons were used and skulls were cracked and various injuries inflicted by the gallant band of constables. This may be considered the proper official wind-up of the week's campaign. What good has been gained by Mr Smith-Barry and his syndicate after all this cruel and wanton outrage is not easily discerned. One hundred and fifty persons have been left homeless, numbers of men and women have been beaten and imprisoned, a priest and a member of parliament have been assaulted and a number of citizens of Youghal have been bludgeoned.

On the following day, it was announced that writs had been served on more than 120 tenants and that the remainder would be processed immediately. 'This means the extermination of 1,600 souls and the depopulation of 50 square miles of country', the *Cork Examiner* editorialised emotively.[69]

The Ponsonby evictions provoked an acrimonious debate in the House of Commons on 21 June, during which J. E. Ellis denounced the Smith-Barry syndicate for interfering in the affairs of the estate: 'It does seem to me an extraordinary thing for an English gentleman or an Irish gentleman to interfere in a matter of this kind for the purpose, which the Hon. Member has avowed, of wreaking vengeance on tenants or of "teaching them a lesson" '. Smith-Barry denied using such expressions, but asserted his right to combine with other landlords to defeat the 'illegal conspiracy called the Plan of Campaign'.[70]

A monster protest meeting was held at Tipperary on the following Sunday, 23 June. In addition to his properties at Fota and Clonakilty in County Cork, Smith-Barry owned most of Tipperary town. In the course of an emotional address for which he was subsequently prosecuted, William O'Brien said: 'The unity of the Irish race in our time is not bounded by parishes or by counties. The Irish cause does not stop at the Galtee mountains. If the landlords of Cork and Tipperary can combine for hate, then I say the tenants of Cork and Tipperary can combine for love'.[71] At O'Brien's instigation, Smith-Barry's Tipperary tenants appointed a deputation and drew up an uncompromising memorial in which they urged him to withdraw from the syndicate and threatened him with a counter-combination if he should refuse. They were now in a position, O'Brien enthused, to meet Smith-Barry and Horace Townsend 'on their own ground and to carry the war into Carthage'.[72] The Tipperary Board of Guardians signalled its approval.[73] More significantly, the redoubtable Dr Croke threw his not inconsiderable archiepiscopal influence behind the tenants' initiative:

> Mr Smith-Barry's intervention in a landlord-tenant dispute that does not immediately concern him proves, beyond all manner of doubt, that he is an aggressive busy-body and a virulent partisan. And indeed, if Mr Smith-Barry can legitimately enter into an alliance with a certain greedy section of the exterminating landlords of south Munster, the aggrieved tenants may fairly claim the benefit and practically have recourse to a similar combination. They may, therefore, it seems to me, become members of any open and constitutional confederacy, forming, in fact, nothing more than a mutual defence association, the object of which would be to make common cause with each other in all agrarian emergencies.[74]

The *Freeman's Journal,* in an adulatory editorial, described Croke's pronouncement as 'the text, the shield, the apologia, and the complete justification of the Smith-Barry Tipperary tenants and all others who may be placed face-to-face with the landlord league'.[75]

Having 'applied the torch and lighted the flame in' County Tipperary, as the police report fancifully phrased it, the agitators now turned their attention to County Cork.[76] A monster protest meeting scheduled for Cork Park on 30 June was banned by the government in the wake of fiery addresses by O'Brien and Keller at Youghal on 27 June and by W. J. Lane at a specially convened meeting of the Cork Harbour Board on the following day. The nationalist press condemned the 'muzzling' of the people by an unashamedly partisan administration.[77]

On Monday morning, 1 July, the *Cork Examiner* editorialised, with great satisfaction, that it would have been easier to keep the tide out of Cork harbour with a pitchfork than to silence the patriotic men of Cork by means of a 'landlord coercionist proclamation'. No less than seven meeting had been held in the city and its environs on the previous day, while William O'Brien addressed a large gathering of Smith-Barry's tenants at Clonakilty. Resolutions condemning the syndicate were passed by bodies as diverse as the Chamber of Commerce, the Young Ireland Society and the United Trades.

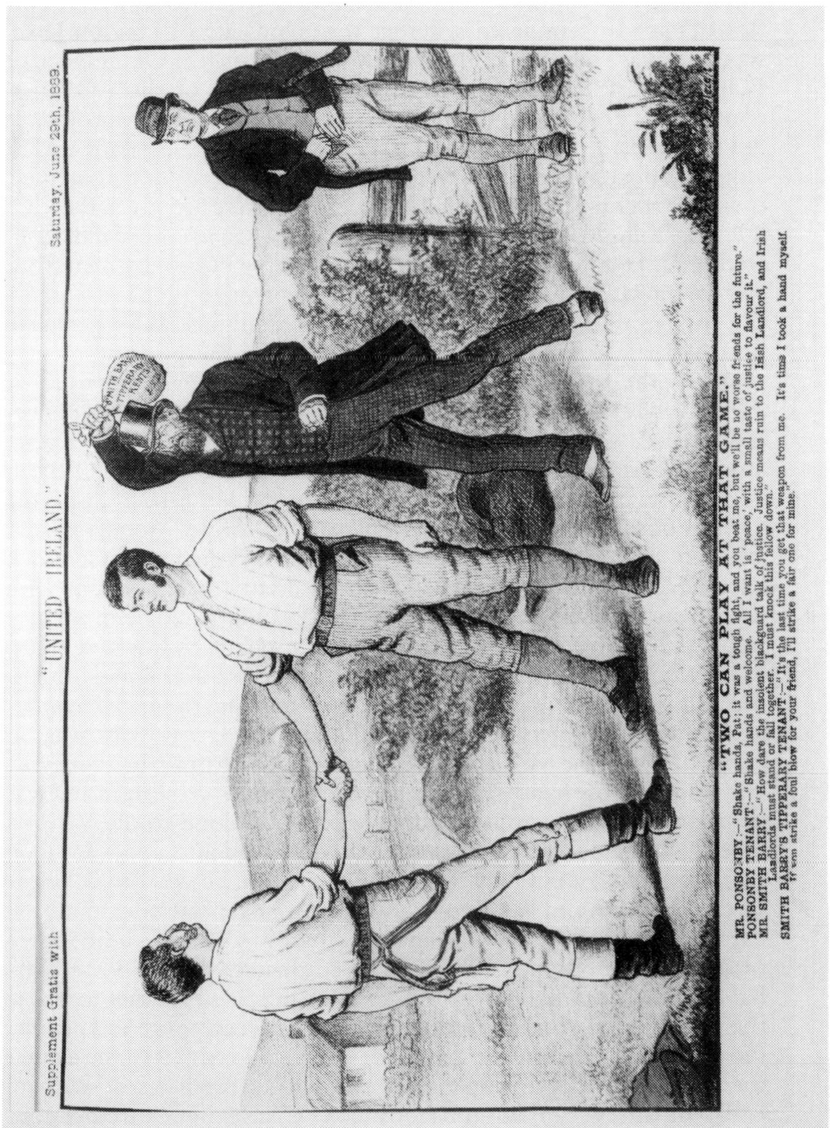
Supplement Gratis with "UNITED IRELAND." Saturday, June 29th, 1889.
"TWO CAN PLAY AT THAT GAME."
MR. PONSONBY.—"Shake hands, Pat; it was a tough fight, and you beat me, but we'll be no worse friends for the future."
PONSONBY TENANT.—"Shake hands and welcome. All I want is 'peace,' with a small taste of justice to flavour it."
MR. SMITH BARRY.—"How dare the insolent blackguard talk of justice. Justice means ruin to the Irish Landlord, and Irish Landlords must stand or fall together. I must knock this fellow down."
SMITH BARRY'S TIPPERARY TENANT.—"It's the last time you get that weapon from me. It's time I took a hand myself. If you strike a foul blow for your friend, I'll strike a fair one for mine."

At Montenotte, Edmund Leamy MP promised that 'combination shall be met by combination'. At the Young Ireland Society's room, P. O'Hea MP wondered whether the 'syndicate of London Jews' would replace the Ponsonby evicted tenants with 'the Orangemen of Belfast, or the scum of towns and the scouring of jails'. O'Brien condemned the Smith-Barry syndicate as an 'atrocious project... devil's work... diabolical conduct... exterminators... insane war of aggression... entered upon by a condemned class rushing upon their own destruction'. He said that Archbishop Croke had given the lead: 'He has sounded the trumpet in Judea and set up the standard in Bethseda. Form one united phalanx behind the Ponsonby men from end to end of Munster'. The Clonakilty tenants resolved to support the citizens of Tipperary in any action they might take to persuade Smith-Barry to withdraw from 'this infamous conspiracy'.[78]

When the train on which O'Brien was a passenger arrived at Albert Quay Station in Cork that Sunday evening, about fifty policemen 'closed up to the carriages and eagerly scanned the passengers, among whom were several clergymen from west Cork'. O'Brien eventually dismounted and was informed by District Inspector Concannon that he had a warrant for his arrest. 'Suddenly and without provocation', the *Cork Examiner* reported, in best mock heroic fashion, 'the police slashed about them in all directions with their truncheons and those of them who had rifles used them as clubs upon the people... They beat the people on the heads and faces, everywhere that a blow would leave its mark'. Several individuals, including reporters from the *Cork Daily Herald* and the *Cork Examiner* and Patrick O'Brien MP, were injured. The latter, who was 'in company with the Misses Wilson' ran forward 'to grasp his namesake and leader by the hand'. He was 'savagely beaten on the head and was brought almost insensible and covered with blood to the Victoria Hotel'. William O'Brien was eventually bundled into a police wagon and taken to Union Quay Station. At 9.55 pm, he was driven to the Great Southern and Western Railway terminus, protesting loudly all the while at 'this *lettre de cachet* proceeding'. In O'Brien's home town of Mallow, thousands of people, accompanied by the local brass band, turned out to meet the train. An unsuccessful attempt to rescue the prisoner led to an ugly confrontation between police and people. Finally, the local curate intervened. 'He could only allay the excitement by telling the people he would shake the hand of William O'Brien for every man in Mallow'. At Charleville, 'a tremendous crowd, wild with excitement' attempted a rescue, to the tune of 'God Save Ireland'. The police opened fire, wounding two. 'The people showed such determination that they dared them to come out and fight them'. When the train finally arrived in Tipperary at 2.30 am, O'Brien was doubly charged under the Crimes Act of 1887 and was released on bail to appear at Tipperary court on 11 July.[79] When the case came to trial it was adjourned to a later date and was subsequently abandoned. Balfour believed that it was contrary to 'the public interest' to proceed with the charge, as it would have the effect of bringing to light all the facts connected with the Ponsonby syndicate. 'It would hardly be fair to the gentlemen who have joined that syndicate', he claimed.[80]

The events of 30 June quickly became notorious. The *Cork Examiner* denounced the actions of the police as 'simply monstrous'. The Mayor and Corporation of Cork were equally censorious. Thomas Sexton condemned the banning of the Cork meeting and the arrest of O'Brien 'as outrageous and atrocious' and moved the adjournment of the House of Commons in protest.[81]

A few days later, the tenants' deputation met Smith-Barry at his Chelsea home. In a prepared statement, he, not unexpectedly, refused to withdraw from the syndicate. He stressed that if a conflict were to ensue with his tenants it would cause 'comparatively slight trouble' to him, whereas it would inevitably mean suffering and misfortune for them.[82] Smith-Barry's response delighted the landlord and tory press. The *Cork Constitution* described it as 'the dignified and firm answer of an honourable man' who refused 'to be intimidated either by the violent language employed by the deputation or the rhodomontade of those who sent them to London'.[83] The *Daily Express* editorialised that every true Irishman would rejoice at Smith-Barry's reply:

> We own to a feeling of the keenest sympathy with behaviour like that of this menaced southern landlord who will not allow himself to be frightened from his path by the menaces of the ignorant, stirred and led on by half-a-dozen raging fanatics . . . The language of his reply . . . is manly, it is bold, it is true'.[84]

Nationalist Ireland, although it could scarcely have expected any other response from Smith-Barry, was outraged. The syndicate was condemned by boards of guardians, town commissioners, corporations and even a grand jury, that of Limerick city. The Cloyne diocesan clergy denounced the 'heartless and ruthless policy' of Smith-Barry and warned of the grave consequences that were likely to result from the continued interference of the syndicate in the affairs of the Ponsonby estate.[85]

Between 8 July and 28 October 1889, Ponsonby took legal proceedings against the remainder of his tenants. At Cork Borough Sessions on 17 July, he was subpoena'd to produce all deeds connected with the estate, especially a deed of trust, executed within the last six months, transferring ownership to Smith-Barry and others. Ponsonby denied the existence of any such document. He had not assigned his interest; he was still the legal owner. The presence of Edward Carson, described by one nationalist newspaper as 'the trusted lawyer of the Castle, the Attorney-General's right-hand man', as his legal adviser excited a good deal of comment. The *Cork Examiner* editorialised that the government was 'in active league' with those who were preventing a settlement of the Ponsonby dispute.[86] At the Wicklow Assizes on 25 July, where more than ninety ejectment decrees from the Ponsonby estate were listed for hearing, the tenants put forward the defence that, as Ponsonby had sold his estate before the writs were served, he had no right to evict them. Ponsonby swore that he had not parted with his property.[87]

A week later, Smith-Barry's Tipperary tenants stated that, while it was the duty of every Irish farmer to prevent the eviction of the Ponsonby tenants by the syndicate, 'a special duty' devolved on them. Accordingly, they agreed to a levy of ten per cent of the poor law valuation of their holdings being raised to support the Ponsonby tenants. To compensate for this 'additional burden' they demanded an abatement of twenty-five per cent in the gale of rent which was then due. According to the *Freeman's Journal,* this marked the completion of the first stage of the movement against the syndicate.[88]

The ensuing conflict between Smith-Barry and his Tipperary tenants—who operated, essentially, a sympathy strike—rapidly degenerated into the most malicious, rancorous and truculent of all the struggles associated with the Plan of Campaign.[89] Nationalist Ireland depicted Smith-Barry as the archest of villains. The *Tipperary Nationalist* denounced him as 'the arrogant aspirant to the title Lord Barry-No-More...unblushing and unscrupulous rack-renter... vainglorious little bashaw'.[90] *United Ireland* described him as 'the pernicious little noodle of a Cork landlord'.[91] W. J. Lane compared him unfavourably with Jack-The-Ripper.[92] Smith-Barry was, as he freely admitted, 'attacked', 'denounced', and 'calumniated' in the radical press of Britain and Ireland. However, he remained philosophical: 'He knew that hard words broke no bones and he was mindful of the old Arab saying: "A man anxious to reach his destination did not stop to throw stones at the dog that barked at him on his way"'. He was convinced that his actions were in the interest 'of true liberty, of law and order and of true freedom'.[93]

8 *The Tenants Defence Association*

> All goes perfectly. Don't have any misgivings or judge hastily on first appearances. When you return you will find the whole country *and party* committed to the struggle of the Plan estates and giving splendid aid. I confess that this will not have come about directly but will be the result of much engineering on both Redmond's part and mine.
> T. P. Gill to John Dillon, 22 November 1889. (TCD Ms.6754, John Dillon Papers).

> The soul and heart of the people rose suddenly.
> Canon Keller to Gill, quoted in Gill to Dillon,
> 22 November 1889.

On his release from prison towards the end of May 1889, William O'Brien found 'things in a very deplorable condition — not a kick in the country and deep depression among the campaigners'. Unless they secured 'money and help', he explained to Dillon, their position would be untenable. O'Brien approached Parnell and convinced him that, if the struggle were to continue, an appeal for £30,000 would have to be launched. Parnell pledged his support, provided it did not bind him to the Plan of Campaign. O'Brien suggested that the Irish leader could best come to their aid by opposing landlord combinations. 'My idea was that he should himself begin the work by attending a convention at Thurles or Cork and this he has all but promised to do'. O'Brien was convinced that, if Parnell openly identified himself with their struggle and inaugurated a series of county conventions, the subsequent parish collections would undoubtedly yield £30,000 'and keep the ranks unbroken'.[1]

On 11 July 1889, the *Freeman's Journal* announced, on Parnell's authority, the immediate formation of a tenants' association to counteract the landlord combination: 'This league will be the official act of the whole Irish party and will shortly be established at a convention'. At an informal meeting in the House of Commons four days later, Parnell called upon the members of the parliamentary party to share with him the responsibility for the new movement.

He impressed upon them the absolute necessity for united action on their part. After a lengthy discussion, it was unanimously resolved:

> That the members of the Irish parliamentary party declare that it is imperatively necessary that the tenant farmers of Ireland should be invited to combine themselves against their attempted extermination by the landlord conspiracy, as instanced by the wanton attack upon the Ponsonby tenants and the threatened application of such aggressive combinations of the landlords to other estates.

A committee of seven—Parnell, O'Brien, Tim Harrington, Thomas Sexton, Justin McCarthy, T. P. O'Connor and T. P. Gill—was appointed to draft a constitution for the new organisation.[2]

A month later, a bitter and disillusioned William O'Brien wrote to Dillon:

> Things have reached a crisis between Parnell and myself. Having started the new Tenants Defence League, with the assurance that it would be used in support of the Plan, he now flatly refuses to take any steps to put life in it. He first allows us to announce that he will throw himself into this fight and now says he cannot see his way to starting any new movement and will not do anything himself.

Parnell refused to attend the Tipperary convention but offered to send Thomas Sexton as his representative. Despite O'Brien's pleas and protestations, the Irish leader remained steadfast. The two parted, after a long and sometimes rancorous argument, on the whole question of agrarian agitation, had brought them to the verge of an open breach. O'Brien determined to 'retort Parnell's policy of silence upon him'; he would not defend the prosecution pending against him, nor would he appeal the anticipated two-months sentence, but would go to prison immediately. Meanwhile, O'Brien would arrange for a few of his most trusted helpers—T. P. Gill, John Redmond and J. J. Clancy, as it transpired—to look after the Plan of Campaign estates. Money from Australia would maintain the agitation until November, by which time he hoped to be free. O'Brien believed that these tactics would force upon Parnell 'the responsibility of getting himself out of the responsibility for the new League' and he was confident that

> Parnell and the rest will be coerced into activity and that the result will be a big fund which cannot possibly be used for any purpose except the support of the Campaigners... Parnell & Co. will be forced to bestir themselves and to recognise that we are not quite

> such helpless victims in their hands as they imagine... It is full time to force him out of a most cruel and infatuated policy of veiled hostility.[3]

On 22 August 1889, O'Brien was convicted at Clonakilty court and spent the next four months in Galway gaol. However, his stratagem to force Parnell to assume some degree of responsibility for the agrarian agitation bore little initial fruit. On 2 September, J. J. Clancy accidentally met the Irish leader in the House of Commons and informed him that a deputation had travelled from Tipperary specifically to speak with him. Parnell declined to see them, 'as he would not undertake the responsibility of advising tenants to give up their interest in their holdings'. Clancy did not inform the members of the deputation that he had met Parnell: 'He did not wish it to be known that he is in London'.[4]

Two months after the initial public announcement concerning the formation of the Tenants Defence League, the Tipperary convention to launch the new movement had still not taken place. On 8 September 1889, John Redmond promised a Tipperary assembly that, before many more weeks had passed, the entire tenantry of Ireland would be banded together in an irresistible combination, 'guided by the unerring wisdom and inspired by the unflinching courage of the great leader of the Irish race, Charles Stewart Parnell'.[5] The problem, as Thomas Sexton confided to Gill, was how to explain the defection of the Irish leader from the inaugural convention. 'After the announcements which have been made in the press and after the expectations which have been excited through the country I think that the absence of Mr Parnell from the initiatory proceedings must create a difficulty which will not be easy to overcome'.[6]

Parnell's nonchalance and his capriciousness regarding the Tenants Defence League, after publicly committing himself and the parliamentary party to it, disenchanted several of his followers. W. J. Lane, who had played a crucial role in the Plan of Campaign on the Ponsonby estate, declined to attend the Tipperary convention. He feared that his presence would lead to another prosecution against him and that a second term of imprisonment would mean certain financial ruin for himself and his family. He had written to Parnell, outlining the extremity of his financial situation and seeking to be reimbursed for the money he had spent attending parliament during the past four years. 'He did not think it worth while even to acknowledge my letter. After ten years work and spending every shilling I had in the cause of which he is leader I was left to sink or swim just as if I had no claim for consideration'. Additionally, Lane believed that the Tipperary tenants should not place themselves in a position analogous to that prevailing on the Ponsonby estate without first receiving some assurance from the leaders of the country that 'they are going to lead and share the fight'. He continued: 'Why don't they come forward and say what they are going to do instead of having everyone asking, "What is going to be done?" Priests and people are asking it every day and we are all like a pack of dummies, unable to give them an articulate reply'. The thoroughly disillusioned MP for East Cork concluded: 'When the Ponsonby fight is over, I retire to private life'.[7]

The month of September was a traumatic one for the organisers of the convention. In the face of Parnell's insouciance, Gill suggested that his name be deleted from the list of committee members. Redmond replied that this 'would do incalculable harm in the country'. To add to their difficulties, Sexton had disappeared and they were absolutely ignorant of his whereabouts. Redmond complained that any further delay 'would be disastrous'. He was determined to hold the convention within a fortnight. 'As to writing to Parnell telling him what we are going to do, it no doubt would be a wise course if you knew his address but I am against delaying for a single day for the purpose of hearing from him'.[8]

For once, Parnell found himself outmanoeuvred by William O'Brien. On 4 October 1889, the Irish leader despatched an open letter to Thomas Sexton, informing him that the state of his health prevented him from attending the Tipperary convention and asking Sexton to deputise for him.[9] Parnell's tacit approval, however reluctantly given, removed the principal stumbling block to the formation of the new tenants' organisation. O'Brien was jubilant. 'The policy of coming to prison and forcing responsibility on the party was completely triumphant', he informed Dillon. 'Parnell would not budge himself but the magnificent attitude of the Tipperary fellows and the resolute pressure of the best of our men forced him to go ahead with the conventions'.[10] Nationalist Ireland heaved a collective sigh of relief, although Parnell's reason for not attending gave rise to a good deal of scepticism. One iconoclast suggested that Parnell preferred 'fornication in Hastings...to a little effort at politics in Ireland'.[11]

At a special meeting of the parliamentary party in the Mansion House on 24 October, the Irish Tenants Defence Association, as it was now called, was formally established and the *Freeman's Journal* editorialised that the ten year old agrarian struggle was entering its final phase.[12] Four days later, the first county convention, attended by 18 MPs, including W. J. Lane, and more than 800 delegates, was held at St Patrick's College, Thurles. The speakers, without exception, were sharply critical of the Smith-Barry syndicate. Thomas Sexton characterised its proceedings as 'mean, vindictive and disgraceful'. John Redmond used the occasion to pay tribute to the Plan of Campaign, describing its history as 'a most extraordinary one'.[13] The series, which continued throughout the winter, was brought to an end by the Kildare county convention on 23 January 1890.[14]

Leading members of the parliamentary party, notably Thomas Sexton, Justin McCarthy, T. P. O'Connor and J. G. Biggar, chaired several of the county conventions. These men had not associated themselves with the Plan of Campaign and their present involvement in the Tenants Defence Association strengthened the public belief that the party was identifying itself more closely with the agrarian struggle than it had done for several years past, a view substantiated by Parnell when, at Nottingham on 17 December 1889, he described the new organisation as 'the greatest, the most powerful and the most peaceful movement that has ever been seen in Ireland or in any other country'.[15] Given Parnell's attitude to the Plan of Campaign, this statement was deeply ironical, since the Tenants Defence Association was essentially a financial rescue operation for the beleagured Plan. Its constitution stressed, as did Parnell in his letter to Sexton on 4 October, that the association was intended to be a defensive combination to combat the landlord syndicate. But the constitution also 'invited' the Irish tenants to contribute to a defence fund, levied at a rate of three pence in the pound on the poor law valuation of their holdings. The £61,000 that was eventually subscribed was used to support tenants evicted from Plan of Campaign estates and to finance the outstanding struggles. The Plan was more or less openly endorsed at the various conventions. At Navan, on 21 November, and at Cavan, a week later, Inspector Alan Bell reported that the speeches

> were in effect a glorification of the Plan of Campaign. Although they did not directly counsel its adoption in the future, the manner in which its success in the past was extolled cannot fail to produce in the minds of the people the belief that it is still a powerful weapon in their hands. No word was said in condemnation of it.[16]

The Tenants Defence Association was strongly supported by the Catholic hierarchy. In the phrase of John K. O'Doherty, Bishop-elect of Derry, it bore 'the imprimatur of the wisest and holiest of the bishops of Ireland'. The organisation had 'the fullest and warmest sympathy' of the Archbishop of Armagh and 'the entire sympathy' of the Bishop of Kilmore, while James Donnelly of Clogher gave it his 'hearty approval'. The Bishop of Dromore indicated that he was in perfect accord with his episcopal brethren throughout the country. Even Dr O'Dwyer allowed his diocesan clergy to attend the Limerick convention after receiving 'an explicit assurance' from its conveners that they would 'abstain from any contravention of the papal decree condemning boycotting and the Plan of Campaign'. The Archbishop of Tuam, John McEvilly, writing to Tobias Kirby at the end of December 1889, insisted that the Tenants Defence Association was essential to prevent the extermination of the Catholic people of Ireland by the landlord syndicate, which he denounced as 'a powerful and opulent confederacy'. He considered it the sacred and solemn duty of a Catholic bishop to support the tenants' organisation. The Bishop of Killala also believed that the aim of the syndicate was to 'exterminate their tenants and clear their estates'. Such a policy was 'reckless and vindictive', according to the Bishop of Cloyne. Dr Donnelly of Clogher reflected the views of his fellow bishops when he emphasised the defensive nature of the Tenants Defence Association, whose aim, he said, was 'to protect by lawful combination the undoubted rights of Irish tenants'. The Bishop of Raphoe attended the Donegal convention and spoke out strongly in favour of the association. The most influential member of the Irish hierarchy, Archbishop W. J. Walsh, addressed the Dublin convention at length. He supported the new movement, he said, because it was non-political and its sole purpose was to defend the tenants and their property. At the Westmeath convention, on the following day, J. J. Clancy MP 'frankly confessed that one of the attractions of the movement was the conviction it produced in him that by smashing Smith-Barry and company they were at the same time smashing Balfour and company'.[17]

If anything, the parish clergy embraced the new organisation with an even greater zeal than their episcopal superiors. They played a crucial financial role, organising and supervising local contributions to the defence fund. Inspector Alan Bell reported: 'The movement has been taken up strongly by the Catholic clergy who have generally assumed the direction and leading parts in local collections, urging their congregations to subscribe not only as a matter of expediency but of charity'.[18] The police reported that, as the collections were generally taken up at the church gates, it was virtually 'a moral impossibility' for an individual to enter without contributing. 'If he did refuse to subscribe, his action was patent to the worshippers passing in and to the loafers around the collectors'. At Kilbane near Killaloe, and at Ardcath and Clonalvy in County Meath, the amount and name of each subscriber was read out from the altar. Fr Pierce of Abbeydorney, County Kerry, took up the subscriptions in the vestry, while, at Scariff, County Clare, Fr McKenna canvassed donations from house to house. The homily preached by Fr McAlpin at Aughmore church, in County Mayo, on 19 January 1890, was by no means exceptional:

> There is no pressure or intimidation to be used but I would like to know the name and see the physiognomy of the man who would stand aside and not come forward and throw in his lot with us in this day of tribulation. Priests, like others, have long memories and as long as I am in this parish neither by word nor act will I do anything for such a man.[19]

The Tenants Defence Association infused new life into the agrarian agitation. For the first time since the autumn of 1886, the whole of nationalist Ireland appeared to support it. Gill informed O'Brien, at the end of May 1890, that there was 'no more lagging spirit in the country or danger of breaks'.[20] Financially, the Tenants Defence Association was a spectacular success. O'Brien had hoped that it would realize £30,000 but, when the fund closed on 25 May 1890, more than twice that amount, some £61,000, had been subscribed.[21] By then, Dillon had returned from the Antipodes and, at a banquet in his honour on 29 May, he disclosed that £33,000 of Australian money had already been remitted.[22] The final figure was expected to exceed £40,000.[23]

The sum of £100,000, collected since the summer of 1889, would, under normal circumstances, comfortably finance the Plan of Campaign for a period of two years. Yet, on the day before the Dillon banquet, Gill wrote to O'Brien:

> Money, money, money is the *one* thing we want now. The thought haunts me. . . so that I go now to the signing of a pile of cheques, to the reading of a pile of applications, to the examination of Ginnell's memos on the interest on our overdraughts with the feelings with which one screws oneself together to go through physical pain.[24]

The building of New Tipperary largely accounts for this incongruity. Of the 355 urban and rural tenants on Smith-Barry's Tipperary estate, only 122 paid their rents up to the last gale day of 1889. The landlord took legal proceedings and, within a year, 154 of his more recalcitrant tenants had been evicted.[25] Several of the shopkeepers attempted to carry on their business by establishing a shopping centre on a neighbouring estate. The scheme caught the imagination of William O'Brien and, on his release from prison in December 1889, he threw himself wholeheartedly into its organisation. At the end of the year, he promised the Smith-Barry tenantry: 'We are in a position absolutely to say that no matter what the struggle costs, no matter how long it lasts, we can, if necessary, match Mr Smith-Barry pound for pound and, if necessary, rebuild this whole Tipperary'.[26]

New Tipperary was opened with a brilliant fanfare on 12 April 1890. Some thirty English and Irish MPs, together with the mayors and public officials of many towns and cities, attended the inaugural banquet.[27] The Under Secretary conceded that New Tipperary was 'a more substantial reality' than he had expected.[28] It was, nonetheless, a quixotic enterprise. Smith-Barry called it 'a cruel and ridiculous fiasco' and, in a letter to Archbishop Croke, made it abundantly clear that he would not surrender to his tenants. He described their action as 'suicidal' and prophesied that New Tipperary would mean their 'absolute ruin'.[29] Logistically, it was probably the single greatest error of the whole Plan of Campaign agitation. Laurence Ginnell, who was in charge of the Plan accounts, stated that £35,000 had been expended on the Tipperary struggle up to the end of January 1891 and a further £5,000 was spent in the ensuing six months.[30] On Good Friday, 1890, T. P. Gill warned William O'Brien that their funds would be exhausted by the autumn and concluded that 'an American expedition will be necessary'.[31]

A resumption of evictions on several Plan of Campaign estates in the early months of 1890 not only added to the Plan's financial difficulties but led directly to a bitter public clash between John Dillon and Dr O'Dwyer, Bishop of Limerick. In January 1890, eviction decrees were obtained against a number of tenants on the estate of John Christopher Delmege, Glensharrold, County Limerick. In mid-May, the receiver of the property offered a clear receipt to August 1889 to all tenants who, before 27 May 1890, paid a year's rent, less thirty per cent. The question of arrears, which averaged three and one-half years' rent, would thus be disposed of. In a circular letter, which was read out to the tenants on 24 May, Dr O'Dwyer appealed to them to accept the receiver's generous offer and thereby save themselves from eviction. Should they act otherwise, he added, they would be 'guilty of reckless folly' and would be unable to convince anyone in the future that they were 'the victims of rapacious landlordism'.[32] During a debate on the Irish etimates in the House of Commons on 11 July, Dillon described the bishop's action as scandalous and denounced his epistle as 'one of the most infamous, cowardly, dastardly letters ever penned by ecclesiastical hand'.[33] Three days later, the letters columns of the *Freeman's Journal* carried O'Dwyer's reply. He deprecated 'Dillon's outburst of outrageous language', claiming that it was 'the conduct of a coward and a dastard to use such language to any man, but most of all to a bishop, behind his back'. He charged Dillon with personal and political cowardice, implicitly accused him and the members of the parliamentary party of financial dishonesty and concluded by roundly condemning boycotting and the Plan of Campaign.[34]

Nationalist opinion was enraged. Under the heading, 'An Episcopal Swashbuckler', the *Roscommon Herald* editorialised: The prelate of Limerick flooded a column and a half of the *Freeman* with a quantity of language which proclaimed its author as a past master in the art of abuse'. The *Sligo Champion* denounced O'Dwyer's letter as 'utterly false and shameless', while the *Westmeath Examiner* proclaimed it to be 'outrageous'.[35] The leader writer of the influential *Freeman's Journal* opined:

> We cannot recall anything of later days, in a public argument between two gentlemen of standing so lofty, more unwarranted and unfair than Dr O'Dwyer's charges of a personal nature against the trusted, the honourable and the fearless John Dillon. To twit with cowardice or dishonesty a man of the character of Mr Dillon ... was so absurd that were the thoughts not penned by a bishop they would be passed unnoticed as the frothings of impotent malignity.[36]

Dillon believed that O'Dwyer had, for the first time, fully revealed himself to the public in his true character and that the bishop's intemperate outburst would benefit the nationalist cause.[37] A monster protest meeting was held in O'Dwyer's own cathedral city of Limerick on 24 August. William O'Brien, Tim Harrington and Dillon, among others, addressed the huge assembly and the latter's speech was, in effect, an unrepentant defence of boycotting.[38] The bishop's reply was immediate and characteristically forthright. Once again, he condemned the 'utterly sinful character' of boycotting and the Plan of Campaign, and stressed, in the wake of the papal rescript, that it was his duty to speak out against them.[39]

Several of O'Dwyer's more conservative episcopal brethren supported his stand. Bishop Coffey of Kerry described the language of the lay Catholic leaders, both inside and outside of parliament, as deplorable and complained to Tobias Kirby in Rome that they seemed determined 'to subvert ecclesiastical authority'.[40] Similar views were expressed by Dr Michael Logue, Archbishop of Armagh, and by Bishop Laurence Gillooly of Elphin.[41] Bishop T. A. O'Callaghan of Cork and Bishop Lynch of Kildare reminded their flocks of the papal condemnation of boycotting and the Plan of Campaign.[42] The more nationalistic members of the hierarchy, however, strongly deprecated O'Dwyer's actions. The two most influential Irish churchmen, the Archbishops of Dublin and Cashel, believed that the Dillon-O'Dwyer contretemps reflected little credit on either man but they were equally convinced, in Croke's phrase, that 'the bishop erred ever so much more grievously than the layman'.[43] In a letter to the Bishop of Elphin, Dr Walsh bitterly criticised O'Dwyer for pursuing 'a crotchety line of his own'.[44]

While this drama was unfolding, the financial crisis facing the Plan of Campaign was becoming ever more severe. Evictions increased the number of dependants on monthly 'doles'; 152 tenants and their families were evicted from the Ponsonby estate alone in the last fortnight of April 1890.[45] Then in the autumn, the potato crop failed throughout the congested districts of the western half of the country. Fr M. B. Kennedy of Meelin, the premier campaigner in north-west Cork, informed Dillon on 20 August: 'A famine will crush the poor in early spring. The potatoe (sic) crop is an utter failure in all this district'. Kennedy believed that they should advise the tenant farmers to refuse to pay any rent until 'they have first bought flour for the sustenance of their little ones'.[46] At Schull, on 14 September, William O'Brien, in a highly-charged address, urged his listeners to avow that

> if ever again the famine pits of Schull and Skibbereen are to be opened it shall not be to receive the corpses of starved peasants famished to death in a land of plenty, but it shall be to receive the foul fiend of Irish landlordism, whose breath is a pestilence, whose footsteps are dogged with famine and desolation and whose crimes there is no pit deep enough to cover or chastise.

O'Brien insisted that the Plan of Campaign was the only remedy available to the distressed Irish peasantry and he advocated its wholesale adoption during the coming winter.[47]

A six-man delegation—John Dillon, William O'Brien, T. P. Gill, T. P. O'Connor, T. D. Sullivan and T. C. Harrington—was due to embark on a fund-raising tour of America early in October 1890. The Chief Secretary was determined to prevent the departure of the two most influential members of the mission. On 18 September, O'Brien was arrested at Roche's Hotel, Glengariffe, and Dillon at his home in Ballybrack, County Dublin. They were arraigned at Tipperary with several others, including the MPs, Paddy O'Brien, Tom Condon and David Sheehy, on charges of conspiracy to induce Smith-Barry's tenants to refuse to pay their rents.[48]

While the conspiracy trials were in progress, an extraordinary meeting of the Irish parliamentary party, attended by sixty-five members, was held in the Mansion House, Dublin, on 6 October, to launch a second appeal under the Tenants Defence Association banner. Parnell was absent, as were the Tipperary defendants and Thomas Sexton, whose mother had just died. The crisis facing the country was described by Justin McCarthy as 'momentous' and by T. P. O'Connor as 'very serious'. Several speakers referred to what John Redmond described as 'the beleaguered estates' and to the wholesale eviction campaigns of the landlords, backed up by the Irish executive, which were intended to smash the Plan of Campaign by starving the tenants into submission. The various speakers insisted that these tenants must be supported. The following motion was adopted:

> That in order to maintain the tenantry of Ireland in just combinations for the defence of their rights and to uphold them against the cruel policy of extermination pursued by combinations of Irish landlords backed up by their richer brethren in England and by the whole strength and resources of a coercion government we feel ourselves reluctantly compelled to appeal to the Irish people to renew their subscriptions to the Tenants Defence Association... and that steps be taken forthwith to open a national fund.[49]

The first contributions were acknowledged in the columns of the *Freeman's Journal* on 8 October 1890, two days after the appeal was launched. On 10 October, Dillon and O'Brien broke bail and spectacularly fled the country. They were due to appear in Tipperary court that morning. The *Freeman's Journal* reported:

> The clerk called out their names. Echo answered. A general titter broke into a peal of laughter. The burlesque was brought to a close amidst general merriment; and the proceedings were dissolved in a comprehensive cachinnation in which even the police joined... They have gone, flown, disappeared. There's a hole in the ballad.[50]

They were smuggled out of Dalkey harbour in a small yacht and, after an eventful voyage, landed at Cherbourg on 15 October, whence they embarked for New York, arriving there on 2 November.

Four days later, they opened their mission at Philadelphia.[51] On Monday, 17 November, the London divorce court delivered its verdict against Mrs O'Shea and Parnell.

9 *Fin de Siècle*

> It is surely a curious instance of the irony of human affairs that this immoral conspiracy [The Plan of Campaign] which has so withstood even the express condemnation of the Holy See should now receive its death blow from the hand of Mrs O'Shea.
>
> *The Tablet,* 19 September 1891.

The Plan of Campaign rapidly collapsed in the wake of the O'Shea divorce case and the subsequent rending of nationalist Ireland into pro- and anti-Parnellite camps. The vital American mission was paralysed from the beginning. On 13 December 1890, Tim Harrington returned to Ireland, while T. P. Gill and William O'Brien embarked for consultations with Parnell in northern France. John Dillon remained in America for another fretful and fruitless month before he too set sail for France, arriving at Le Havre on 18 January 1891. A month previously, he had written to inform Harrington that they were almost at the end of their resources.[1] The latter replied, on New Year's Eve, that Irish affairs were in 'a sad fix' and that there was bound to be 'a frightful collapse' unless American aid arrived immediately.[2] Laurence Ginnell stated that they were supporting 1,600 families and that the Plan of Campaign bank accounts were in 'a desperate condition'.[3] According to James F. X. O'Brien, they were overdrawn by more than £10,000. In addition, £2,300 was required for grants to evicted tenants before mid-January and there were pressing demands for another £1,700.[4]

After the collapse of the Boulogne negotiations, Dillon and O'Brien crossed to Folkestone, where they were arrested and transferred to Galway gaol, there to serve the two concurrent terms of six months' imprisonment imposed on each of them after their flight from Tipperary in October 1890. Thus, at a crucial time, the Plan of Campaign was deprived of the guidance of its principal leaders. In addition, the split in the parliamentary party was reflected in the ranks of the campaigners. Several leading advocates of the agitation sided with Parnell, including the Redmond brothers, Dr Kenny, J. J. Clancy, Luke Hayden and John O'Connor.

The tribulations in the nationalist camp demoralised the tenants. The failure of the American mission meant that grants were either delayed or were not paid at all and there were several instances of cheques being dishonoured.[5] The £8,000 freed from the 'Paris funds'[6] for the relief of evicted tenants was largely used to placate anxious bank managers.[7] A police report on the Olphert estate in County Donegal, drawn up on 10 January 1891, noted that the question of funds was 'vital and pressing'. If the Tenants Defence Fund, which had been launched in the previous October, proved to be a failure, the report concluded, 'the evicted tenants here, as elsewhere, will be at the mercy of the landlord absolutely'.[8] If 'the wounded soldiers of the land war' faced a problematical future, tenants under threat of eviction were determined to avoid a similar fate. The result was a noticeable improvement in the payment of rent. Divisional Commissioner Byrne claimed that the Parnellite split had 'a most important bearing on the rent question', while Alan Bell, of the Special Branch, estimated that it was worth 'at least ten per cent to the landlords in their rents'.[9]

At the end of January 1891, George Shaw Lefevre, who had championed the Irish tenants' cause on several previous occasions, moved in the House of Commons that the government appoint a board of arbitration to settle outstanding Plan of Campaign disputes.[10] This proposal was not a new one. In October 1887, Archbishop W. J. Walsh suggested that George F. Brooke and his County Wexford tenants should submit their dispute to arbitration. He proposed that each side appoint two arbiters, with power, if necessary, to call in an umpire. Their adjudication should not be based on any capricious estimate of their own but on the judicial rents then being fixed throughout County Wexford by the land courts. Walsh's proposal did not meet with any success.[11]

On the occasion of William O'Brien's trial at Killarney, some eighteen months later, County Court Judge Curran suggested that the Plan of Campaign on the estate of Lord Kenmare should be submitted to the arbitration of two English MPs, with Curran himself acting as umpire.[12] The Chief Secretary was appalled, describing the judge's proposal as 'preposterous and indefensible'.[13] The trustees of the estate, in rejecting the scheme, obviously held a similar opinion.

A few days later, at the beginning of May 1889, the agrarian conflict on Colonel Vandeleur's County Clare property was brought to a successful conclusion by means of arbitration and, thus encouraged, Archbishop Walsh proposed in the *Times* that all outstanding Plan of

Campaign disputes be similarly dealt with.[14] There was little positive response, the landlords in general were suspicious, while the Chief Secretary was implacably opposed to the idea. By definition, arbitration meant compromise and, in Balfour's opinion, this smacked of a victory for the Plan of Campaign. 'Any compromise with it', he claimed, 'which leaves its prestige unaffected in the eyes of the people is practically a victory and every victory means a recrudescence of the Plan in some new and hitherto untroubled district'. But, he warned the Under Secretary, arbitration sounded so plausible that it should not be rejected 'roughly or unnecessarily'.[15]

After the Parnellite split, the Chief Secretary displayed no such delicacy of feeling. Graphically and unambiguously, he rejected Shaw Lefevre's motion in the House of Commons. Balfour asserted that he would never counsel any landowner in Ireland to submit to the Plan of Campaign and if he were an Irish landlord he would beg his bread before surrendering to it. He said he would be 'insane' to introduce a measure of compulsory arbitration whose sole effect would be 'to bolster up the tottering Plan'.[16] The *Times* concurred, editorialising that arbitration would be 'a sure and certain method of propagating and perpetuating the conspiracy throughout Ireland'. Landlords were urged, as a duty 'to themselves, to their country and to their order,' to follow Balfour's advice and not to compromise with the 'policy of public plunder'.[17]

The owners of campaigned estates proved to be singularly uncompromising. Many seemed determined to avenge themselves on their tenants for the pressures they had endured and the humiliations they had suffered over the previous five years. On 23 January 1891, John Fitzgibbon, one of the leading campaigners in the west of Ireland, informed John Dillon that Lord De Freyne was 'acting like a tiger. He appears thirsty for vengeance on the people that brought him to his senses before'. Ten days later, Fitzgibbon reported that De Freyne was determined 'to smash at any cost the tenants' combination by refusing every reasonable concession'. If he succeeded, Dillon's informant concluded, 'the good work that has been done on the neighbouring estates during the past five years will receive a damaging blow'. Fitzgibbon claimed that it would 'put an end to all future combinations for many a day and will give new life to tyrannical landlordism'.[18] Fitzgibbon had 'guaranteed' the De Freyne tenantry, at a public meeting on 6 January 1891, that 'no matter what occurred, they would be sustained to the last. An honourable settlement, or no settlement, must be got', he said.[19] By 5 February, the despairing and panic-stricken tenants had surrendered on De Freyne's terms.[20] The police believed that this settlement and the one on Lord Dillon's neighbouring property played a major role in persuading tenants on smaller estates throughout the west to come to terms with their landlords.[21]

The Plan of Campaign entered its terminal phase in April 1891. The collapse of the agitation on the Delmege estate in County Limerick was regarded by the police as a breakthrough. The tenants asked Dr O'Dwyer to intercede with the Court of Chancery on their behalf. Originally, they had demanded an all-round abatement of fifty-seven per cent and, on several previous occasions, had rejected the court's proposals to reduce the rents by thirty per cent. When this offer was repeated, through the Bishop of Limerick, in April 1891, the tenants were glad to accept. Subsequently, John O'Connor, John O'Connell and James Downey denied a report in the *Limerick Leader* that the evicted tenants on the Delmege estate had been well cared for; that they had been in receipt of grants from the time of their eviction and that such grants had been paid promptly. They stated that, when they were originally 'induced' to adopt the Plan of Campaign, they were promised that they would 'want for nothing' and if they were evicted they would 'be gentlemen going back' to their homes and farms. Instead of which, they continued, 'our houses are ruined, our farms are like a wilderness and our pockets are empty'. In conclusion, they tendered their heartfelt thanks to Dr O'Dwyer and to Fr Walsh for interceding on their behalf and recorded their 'everlasting hatred' for Edward Fitzgerald, the man responsible for the Plan of Campaign in the Glensharrold district.[22] On 1 May, District Inspector Jones of the Cork Special Branch reported:

> One of the greatest triumphs over the anarchists during the league agitation has been the collapse of the Plan of Campaign during the past month. No matter what the anti-Parnellites may write or say the Plan is a thing of the past. Its victims are cursing the day they were induced to adopt it.

Colonel Alfred Turner claimed that 'the Plan of Campaign fraud' had collapsed in his division and that the nationalist split had done more for the cause of law and order in five months than he had been able to do in five years of arduous work.[23]

In mid-May, it was reported that the evicted tenants on The O'Grady's County Limerick estate had not received any grants for more than two months. J. F. X. O'Brien advised them to settle as there were no funds available to pay them.[24] Parnell and John Redmond had already publicly stated that the Plan of Campaign could no longer be sustained and strongly urged tenants to come to terms with their landlords.[25] The financial situation continued to deteriorate during the summer and by the end of July the funds of both Parnellites and anti-Parnellites were completely exhausted. From the beginning of August, the National League was compelled to discontinue its support of the evicted tenants entirely,[26] while National Federation 'doles' were paid out once every two months, instead of monthly as heretofore. The latter organisation had spent £30,000 between January and July 1891 and was compelled to launch a public appeal for funds in September.[27]

By mid-1891, desperate tenants had begun to make extremely unfavourable settlements and to accept terms which contemporaries regarded as ruinous. A Tipperary tenants' deputation met their landlord, Arthur Hugh Smith-Barry, at Chester on 25 May. They had come to accept that it was 'hopeless' to prolong the struggle any further. 'Instead of ruining him', they said, 'they were ruining themselves'. Smith-Barry's terms represented a comprehensive defeat for the tenants. The members of the deputation admitted that they were 'not generous to the tenants', but insisted that they were the best they could obtain under the circumstances.[28]

At the end of September, Smith-Barry rejected as 'too ridiculous for discussion' a proposal that the dispute on the Ponsonby estate be submitted to arbitration.[29] A few weeks later, he announced in the *Cork Examiner* that he was prepared to negotiate with the Ponsonby tenants under the purchase clauses of the recently enacted government land measure. At the beginning of November, he offered to sell at seventeen and nineteen years' purchase of the non-judicial and the judicial rents respectively. In addition, the tenants were to pay from twelve to eighteen months' rent in lieu of arrears. They, in their turn made a counter-offer which Smith-Barry rejected. According to Canon Keller, the tenants were in a very difficult position:

> We could not promise anything like a regular continuance of the grants the tenants had been receiving...I told them that if the money was there they would get it but I could not promise that the regular monthly grants would be sent to them or that they would not stop altogether...I said, "you have to decide on your own responsibility but my advice to you is to try and hold on; have patience for a little while and do not sign terms you admit to me to be perfectly impossible".

The six-months time limit on the purchase clauses of the 1891 Land Act was due to expire on 5 February 1892. Between 2 and 5 February, about a hundred tenants purchased their holdings on Smith-Barry's terms. Keller described this development as 'absolutely sad and distressing'. All but two of the tenants 'believed the terms to be impossible... but the question always was, "What are we to do?"'[30]

At the end of January 1892, five of The O'Grady's tenants agreed to pay one and a half years' rent in lieu of arrears and costs and to buy their holdings at eighteen years' purchase of the judicial rents. According to one local man, the conditions were not too bad considering those on which the Ponsonby tenants were about to settle.[31] In March, more than one hundred of Wybrant Olphert's County Donegal tenants accepted his offer to reinstate each of them on payment of three years' arrears of rent up to 1 November 1891 and costs of thirty shillings. Fr John Boyle regarded these tenants 'as doomed men. Their condition is extremely bad. I know for a fact they are sunk in the shops, that the shopkeepers have refused them credit and they are pressed even now for another year's rent'.[32]

In January 1892, the National Federation was supporting the families of 1,500 evicted tenants.[33] By the end of the year, Dillon informed the Evicted Tenants Commission, the number had fallen to 884 and seventeen Plan of Campaign disputes still remained unresolved.[34] During the succeeding years, settlements continued to be arrived at and, by April 1905, peace had been restored to all but seven Plan estates.[35] Over the next two years, the Brooke, Byrne, Lansdowne, Massereene and Ponsonby properties were either sold or were in the process of being disposed of to the tenants.[36] In June 1906, Mr H. H. Lewis informed the Land Commission that he would no longer oppose the restoration of the evicted tenants to his property. Purchase negotiations were opened but another two years were to elapse before they were satisfactorily concluded. The 441-acre estate was sold for £6,920. The seventeen evicted tenants took up possession of their newly acquired farms on 4 January 1909.[37] In 1913, Lord Clanricarde was compelled by a decision of the Land Court to sell his estate to the Congested Districts Board, under the compulsory purchase provisions of the 1909 Land Act. This eccentric, absentee landlord—whom William O'Brien once described as 'one of the strangest monsters in recorded history... a creature so abnormal that the first duty of a civilized state would have been to put himself and his affairs in sequestration'—received more than £238,000 in compensation.[38]

Conclusion

In the spring of 1894, Michael Davitt claimed that the Plan of Campaign was lauded far too much by its authors and credited with a great deal more than it ever did or could possibly accomplish.[1] The claims made for the Plan by some of its proponents were indeed extravagant. In Adelaide on 12 April 1889, John Dillon asserted that the agitation

> had turned back the crowbar and the evictor from many thousands of Irish homes...It has broken the power of landlordism in Ireland and has inflicted on the landlords a greater check than any other policy of combination ever started for the relief of the people of Ireland, but what I most value...is that it has saved the people of Ireland from the Ribbon lodge and the Whiteboy association...Where the Plan of Campaign is planted, there crime is not to be found.[2]

To suggest that the Plan curbed evictions and reduced the incidence of agrarian crime is to distort the facts. Prior to the passing of the Land Act of. 1887, Sir Michael Hicks Beach had constrained the landlords' evicting zeal. From the summer of 1887 onwards, the ever-spiralling number of ejectments from Plan of Campaign estates, and the consequent drain on limited resources, was the primary concern of the Plan's leadership. The reduction in agrarian crime is largely explained by the Balfour administration's stringent use of coercion and by the almost universal use of boycotting, a moral sanction whose very nature lessened the need for physical violence.

On the positive side, the Plan was successful on most of the estates where it was introduced and secured sizeable rent reductions for many tenants. In addition, the agitation had a considerable, if incalculable, indirect influence on rent movements, many landlords conceding their tenants' demands on being threatened with the Plan or on seeing it take root in the locality. W. J. Lane claimed that the agitation on the Ponsonby estate would serve as 'a scarecrow' to neighbouring landlords.[3] William O'Brien said that the Plan was like having 'a good mastiff on the premises. The very bark of him is enough to warn off marauding landlords'.[4]

The Plan of Campaign was not a national movement. While it was more extensive than has hitherto been realised, it directly affected a mere one per cent of Irish estates.[5] Nor could the expenditure on the agitation be considered inordinate. Between October 1886 and December 1892, John Dillon informed the Evicted Tenants Commission, a total of £234,000 was subscribed to the Plan. The Irish public contributed £129,000, £62,000 came from Australia and America, £1,500 from Britain, while the tenants themselves subscribed £42,000 to the 'war chest', £30,000 of which was subsequently returned to them. The balance was disposed of as follows: £125,000 was paid in monthly allowances to evicted tenants, £51,000 was spent on housing, legal costs accounted for £11,500 and the remainder was spent on miscellaneous items, such as travelling, hotel accommodation, grants to aid settlements, etc.[6] An independent police inquiry into the Plan's finances, in May 1892, substantially corroborated Dillon's figures.[7]

The hysterical reaction to the Plan of Campaign, which seems disproportionate to its scale, is due, in no small measure, to the implications of the agitation. Although it was rooted in the economic malaise of the mid-1880s, the Plan was always more than an attempt to secure temporary rent reductions. Most of its victims appeared to be carefully chosen and a landlord's known financial vulnerability was the overriding criterion. The loss of their rents was critical for many landowners. In March 1893, the land agent, William Rochfort, claimed that the Marquis of Lansdowne had spent 'not less than £10,000' in combating the agitation.[8] The Plan cost The O'Grady £6,416, exclusive of the considerable losses sustained in the working of his 400-acre demesne, which was boycotted down to 1891. The Anti-Plan of Campaign Association paid his legal costs of £1,436 and provided the capital to stock the evicted farms, for which it subsequently received £1,298 as a percentage of the profits. The estate was very heavily mortgaged and, according to the agent, would have passed out of The O'Grady's hands had not several of the mortgagees, who were related to him, reduced the annual interest rate to three per cent and waived one year's payment entirely. The Dowager Madam O'Grady reduced her jointure from £500 to £100 and Madam Elizabeth O'Grady maintained the household at Kilballyowen from an independent income of her own.[9] The arrears of interest on the Countess of Kingston's massive £236,000 mortgage increased from £3,000 at the end of 1886 to more than £13,000 a year later.[10] In January 1888 the mortgagee placed the estate under the control of the Court of Chancery. Receivers were also appointed to the properties of the Earl of Granard, Arthur Langford, J. C. Delmege and W. N. Leader during the Plan of Campaign years.[11]

Bankrupt landlords would have little option but to sell their estates on punitive terms to their tenants. The leaders of the agitation made no secret of their ambition to abolish the institution of landlordism, which was odious to the majority of Irish people. Additionally, and perhaps more significantly, the landlords were seen, and largely regarded themselves, as part of the English garrison in Ireland and, as such, were a considerable barrier to the establishment of a Home Rule parliament in College Green. The highly-charged political summer of 1886 had unleashed explosive passions on both sides of the Irish Sea and it was inevitable that the Plan of Campaign should become imbued with some of the ideology and assume many of the excesses of the Home Rule debates. The opponents of the agitation were convinced that it was politically motivated. One unionist organisation insisted that the Plan was 'nothing more or less than a side stroke of the Separatist attack'. Sir Michael Hicks Beach claimed that its aim was the 'eventual extirpation of... English rule in Ireland,' while the Prime Minister argued that the Plan would ultimately lead to 'the dismemberment of the Empire'.[12]

Accordingly, the agitation was opposed by all those who regarded themselves as upholders of the political and social status quo. The refusal to pay rent was a direct subversion of what was seen as the sacrosanctity of contract, as enshrined in the Land Act of 1881 and, as such, was denounced as illegal and morally indefensible. On the nationalist side, papal condemnation of the Plan of Campaign and of boycotting in the early summer of 1888 was rebutted as unwarranted meddling in Irish political affairs. The dazzling display of casuistry, chiefly inspired by Archbishop Croke and John Dillon, that greeted the Pope's action, provided a further indication, if one were needed, that on questions of morality, especially political morality, an infinity of angels could, indeed, dance on the head of a pin. When the Prime Minister made a swingeing attack on the Plan of Campaign at the City Conservative Club, London, on 8 December 1886, he drew a retort of withering finality from the Kilteely National League:

> That we recognise in the truculent speech of Lord Salisbury... yet another incontestable proof that our hereditary enemies, the Tories, can boast, as the Bourbons of old, that if they have learned nothing, they have forgotten nothing; and we take this opportunity

> of saying to the noble lord that now, as on all former occasions, we decline to take our lessons in morality from any English source, particularly a political one.[13]

The land legislation of 1887 empowered the courts to revise all judicial rents fixed between 1881 and 1886 and, thus, the Salisbury government, which placed so much store on property rights and sanctity of contract, effectively overturned the Land Act of 1881. The concept of dual ownership was clearly shown to be inoperable and peasant proprietorship was increasingly seen as the only possible solution to the Irish land question. Additionally, the Plan of Campaign had increased the tenants' political awareness and further alienated them from the mainly Protestant landed class, whose political and social isolation was now almost complete.

Standish O'Grady apostrophised the Irish landlords in 1886: 'Your career is like some uncouth epic begun by a true poet, continued by a newspaper man, and ended by a buffoon; heroic verse, followed by prose, and closed in a disgusting farce. Then, *plaudite* and *exeunt omnes.*'[14] To complete the theatrical allusion, the Plan of Campaign presaged the final curtain.

APPENDIX I

A PLAN OF CAMPAIGN*

A Memo. For The Country

Except in a few districts of Ireland, evictions have not been as numerous during the past two months as in the earlier portions of the year. This is simply the annual breathing time of the crowbar brigade. The 'long vacation' in the Superior Courts has, however, now drawn to a close, and already the County Courts have resumed their quarterly sittings. Ejectment processes have been issued wholesale, writs of the Superior Courts will soon follow, and again the fell work of destruction will proceed. Irish homes sanctified by many happy memories will be unroofed; Irish families will be scattered. One who has borne his part in the struggle of the past half dozen years, who has seen almost every phase of the agitation, thinks it his duty at the present critical moment to offer to the Irish tenantry and their friends the following suggestions as the result of his experience. Present rents, speaking roundly, are impossible. That the landlords will press for them let the rejection of Mr Parnell's Bill testify. A fight during the coming winter is therefore inevitable, and it behoves the Irish tenantry to fight with a skill begotten of experience. The first question they have to consider is

HOW TO MEET THE NOVEMBER DEMAND

In a few weeks at most the agents will issue invitations from the rent-office. There should not be an estate in Ireland where the tenants would not by that time have their minds fully made up as to the course they intend to pursue. To delay action until the gale-day means to go into the struggle handicapped. Should combinations be formed on the lines of branches of the National League or merely by estates? I say by estates decidedly. Let branches of the National League, if they will, take the initiative in getting the tenantry on each estate to meet one another. But it should be distinctly understood that the action or resolution of one estate was not to bind any other, and the tenantry on every estate should be free to decide upon their own course. When they are assembled together, if the priest be not

*Published in *United Ireland* on 23 October 1886 and re-issued, as a special supplement, on 20 November 1886.

with them, let them appoint an intelligent and sturdy mem ser of their body as chairman, and after consulting, decide by resolution on the amount of abatement they will demand.

A committee consisting, say, of six and the chairman, should then be elected, to be called the Managing Committee, and to take charge of the half-year's rent of each tenant should the landlord refuse it.

Everyone present should pledge himself (1) to abide by the decision of the majority; (2) to hold no communication with the landlord or any of his agents, except in presence of the body of the tenantry; (3) to accept no settlement for himself which is not given to every tenant on the estate.

Should any tenants be excepted? Question is likely to be raised as to large holders. It should be remembered, however, that the large tenant plays for high stakes, and there is no reason why he should not throw in his lot with the rest. Holders of town-parks who are shopkeepers have a stronger claim to exemption, for a judgement against them may mean ruin. But no case for exemption arises at this stage until it be known how the landlord will proceed. In any published report of the meeting the names of the committee should not be given.

On the gale-day the tenantry should proceed to the rent-office in a body. If the agent refuses to see them in a body they should on no account confer with him individually, but depute the chairman to act as their spokesman and acquaint him of the reduction which they require.

No offer to accept the rent "on account" should be agreed to. Should the agent refuse—then every tenant must hand to the Managing Committee the half year's rent which he tendered to the agent. To prevent any attempt at a garnishee this money should be deposited by the Managing Committee with some reliable person whose name would not be known to any but the members of the committee. This may be called the Estate Fund, and it should be absolutely at the disposal of the Managing Committee for the purposes of the fight. Broken tenants who are unable to contribute the reduced half year's rent should at least contribute the percentage demanded from the landlord—that is, the difference between the rent demanded and that which the tenantry offer to pay. A broken tenant is not likely to be among the first proceeded against, and no risk is incurred by the general body in taking him on these terms. Thus practically a half-year's rent of the estate is put together to fight the landlord with. This is a fund which, if properly utilised, will reduce to reason any landlord in Ireland.

HOW SHOULD THE FUND BE EMPLOYED?

The answer to this question must to some extent depend upon the course the landlord will pursue; but in general I should say it must be devoted to the support of the tenants who are dispossessed either by sale or ejectment.

It should be distributed by the committee to each evicted tenant in the proportion of his contribution to the fund.

A half-year's rent is supposed to maintain a tenant for half a year, and based upon this calculation a tenant who funded say £50 would be entitled when evicted to receive £2 per week.

But not one penny should go in law-costs. This should be made an absolute rule, for there is no principle in the whole agitation more grossly misunderstood. I have known combinations where the tenants agreed to pay the law-costs for any man proceeded against, and all the while they never seemed to realize that it was the landlord's law-costs they were paying, and that they were helping to defeat their cause rather than serving it. Incidents undoubtedly may arise in the course of the struggle where the expenditure of some small sum in legal defence would be judicious, such as defending a tenant who takes possession of his home. But these are exceptional cases, and can be easily provided for. The "law-costs" which must be guarded against are the costs of attorneys' letters, writs, and judgements incurred by the landlord. To pay these means to arm your enemy for the quarrel and furnish him with provisions to boot. In a determined fight there are no "law-costs" on the side of the tenantry, and they should remain out for ever rather than pay those which the landlord incurs in fleecing them.

Grants and expenses should all come out of common fund. Everyone settling before majority have agreed to accept settlement should forfeit his contribution. When the tenants decide upon settling, the balance in hands should be divided among them in proportion to the amount they funded.

THE ASSISTANCE OF THE NATIONAL LEAGUE.

To inspire confidence among the tenants, the National League should guarantee:-

1st. That in the event of the trustee with whom the Estate Fund was lodged proving dishonest, the money would be made good to the tenants, and grants in proportion to their contributions given to them just as if their own fund had not disappeared.

2nd. That when the Estate Fund of the tenants has been expended, or so diminished as not to be able to meet the grants required, then the grants would be continued on the same scale by the National League.

3rd. That this grant would be continued as long as the struggle lasted and the majority of the tenants held out.

THE LANDLORD'S REMEDIES.

Before referring in detail to the different remedies open to the landlord, it is first necessary to caution the tenantry against some artifices to which it is not uncommon for the landlords to have recourse. Rumours are industriously sent round after the gale-day that a certain proportion of the tenants have secretly paid their rent. Instances have occurred where bailiffs, by trying to get into conversation with some of the tenants at public places or making an ostentatious call at their houses, try to spread the belief that certain men are breaking away from the combination; and to push the deceit still farther, only every alternate tenant in a town-land is sometimes served with the first attorney's letter. I need only mention that these artifices are practised to put tenants off their guard. So long as the main body, the majority in fact, of the tenants hold out they need give little heed to rumours about back-sliding. Those who do back-slide in such a case invariably come off the worst of the whole body.

EJECTMENT

is the most common of the landlord's remedies. The procedure, unfortunately, is too common to need much explanation. When a landlord intends to seize cattle or sell the interest in the holding, the writ or process states that the plaintiff's claim is for £. s. d. (setting forth the amount of money due); when on the other hand the claim is for the recovery of possession, etc., then he is proceeding by ejectment. In ejectment the tenant is allowed six months after eviction to redeem, the landlord being bound to account for the crops, etc., if not left to the tenant.

Every legal and constitutional obstacle which could oppose or delay eviction should be had recourse to.

Every hour by which the sheriff is delayed in one eviction gives another brother tenant so much more grace. There are only 310 days in the sheriff's year, and he must do all the evictions in a whole county within the time.

If, after eviction, a tenant is readmitted as caretaker, he should go in, but never upon the understanding that he would care any other farm but his own.

Should the tenant not be readmitted shelter must be procured for him immediately by the Managing Committee, and then if necessary, a day appointed when all would assemble to build him a hut on some spot convenient to the farm where the landlord could not disturb him. Wooden huts such as those supplied by the Land League waste too much of the funds, and become valueless when the tenant is readmitted.

Sometimes it happens, that when a landlord evicts he burns, or otherwise demolishes, the home of the evicted family. Here the Managing Committee would do well to consider, if only little rent is due, whether it would be wise to redeem and take an action against the landlord to compel him to repair the injury.

Any labourer evicted in consequence of the strike should be carefully seen to by the committee, and provided with suitable shelter and employment.

SALE.

This is the resort of the landlord when he proceeds by writ or process as an ordinary creditor. From eight to twelve days are allowed after service of the writ before judgement can be marked. The sheriff may seize cattle if he finds them on the farm, or he may seize and sell the tenant's interest in the farm. A tenant who has his mind made up for the fight will have his cattle turned into money before the judgement comes on. Every tenant who neglects to dispose of them is preparing himself to accept the landlord's terms, for he will not wish to see the emergency-men profit by taking his cattle at some nominal price, and if he buys he is in reality handing the landlord the amount of his demand. Sale of a farm is not of so much consequence. Every farm sold in this manner during the agitation either has come or is bound to come back to its owner even on better terms than he first held it.

But if a man has a very valuable interest in his farm he can place it beyond the sheriff's power by mortgaging it to some one to whom he owes money. Mortgage effected thus for a bona fide debt or consideration, bars the sheriff's power of conveyance.

At a sale, if the landlord or emergency-men be represented, the cattle should not be allowed to go at a nominal sum. They should be run up to their price, and, if possible, left in the hands of the emergency-men at full price. It must be borne in mind that if the full price be not realized, the sheriff could seize again for balance.

In bidding for a farm it should also be run to amount of debt, but by a man of straw, or some one who, if it were knocked down, would ask the sheriff for time to pay. By making the landlord's bidder run it up to the amount of debt and costs, and leaving it on his hands, the sheriff cannot follow the tenant further.

No auction fees should be allowed.

A farm held on a lease for a life or lives, any one of which is extant, cannot be sold by the sheriff.

After sale a tenant is still in possession of holding until a fresh writ is served and a judgement for title marked against him.

All this involves the landlord in fresh costs.

The eviction may then follow, and my observations in case of eviction on ejectment apply here.

DISTRESS,

another of the landlord's remedies, cannot be resorted to for more than one year's rent. Few landlords can have recourse to this without exposing themselves to actions. The chief points to attend to are:- That distress must be made by landlord or known agent, or bailiff authorized by warrant signed by the landlord or known agent; that particulars of distress be served; seizure on Sunday is unlawful; seizure before sunrise or after sunset is unlawful; or for any rent due more than one year. Distress is illegal if growing crops be seized; or the implements of a man's trade; and if other property be on farm to answer landlord's demand, it is illegal to seize beasts of the plough, sheep, or implements of husbandry necessary for the cultivation of the land. These points should be carefully watched when landlord has recourse to Distress.

BANKRUPTCY PROCEEDINGS

are too costly a machinery for general use, and no landlord is likely to have recourse to them except in dealing with a holder of a town-park, or someone who has an industry in addition to the farm in dispute. As I hinted already, such a tenant might be exempted by the general body from any action which involved the marking of judgement against him. But he should at least remain in the strike to the extent of leaving his money in the common fund until the struggle be over.

It is unnecessary to add that landlords and their partizans on the magisterial bench, and among crown officials, will do all in their power to twist the operation of the law so as to harass the tenants.

A tenant taking possession of his home to shelter his family from the severity of the winter, is not likely to escape. A summons for trespass must be preceded by a warning to the tenant if he be found in possession. I have known a case where the father complied with this warning, and on the bailiff's next visit the mother only was found, and she complied. Next time the eldest daughter only was in possession, and so on through the length of a long family such as an evicted tenant nearly always has. A goodly time had been saved before the father's turn came again. He was fined, and went to jail. Prison then lost its terrors for him. When he came out he stuck boldly to his home, and he soon won the victory which rewards determination.

PUBLIC SYMPATHY.

The fullest publicity should be given to evictions, and every effort made to enlist public sympathy.

That the farms thus unjustly evicted will be left severely alone, and everyone who aids the evictors shunned, is scarcely necessary to say.

But the man who tries boycotting for a personal purpose is a worse enemy than the evicting landlord, and should be expelled from any branch of the League or combination of tenants.

It has often surprised me that tenants waste their energy in puny efforts to boycott a few bailiffs or emergency-men.

On estates where some tenants have been unjustly evicted, the others are paying their rents with punctuality, while they expect publicans and shopkeepers and others to boycott police or emergency-men.

No landlord should get one penny rent anywhere, or on any part of his estate, wherever situated, so long as he has one tenant unjustly evicted.

This policy strikes not only at the landlord, but the whole ungodly crew of agents, attorneys, and bum-bailiffs.

Tenants should be the first to show their sympathy with one another, and prompt publicity should be given to every eviction, that the tenants of the evictor wherever he holds property may show their sympathy.

Such a policy indicates a fight which has no half-heartedness about it, and it is the only fight which will win.

Map A: Distribution of Plan of Campaign Estates.

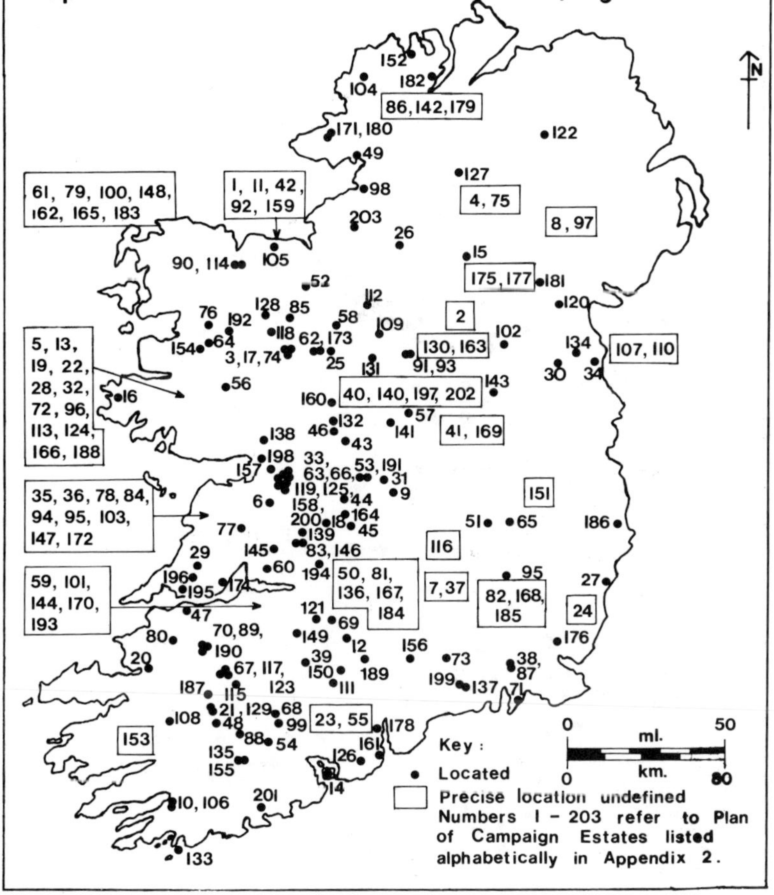

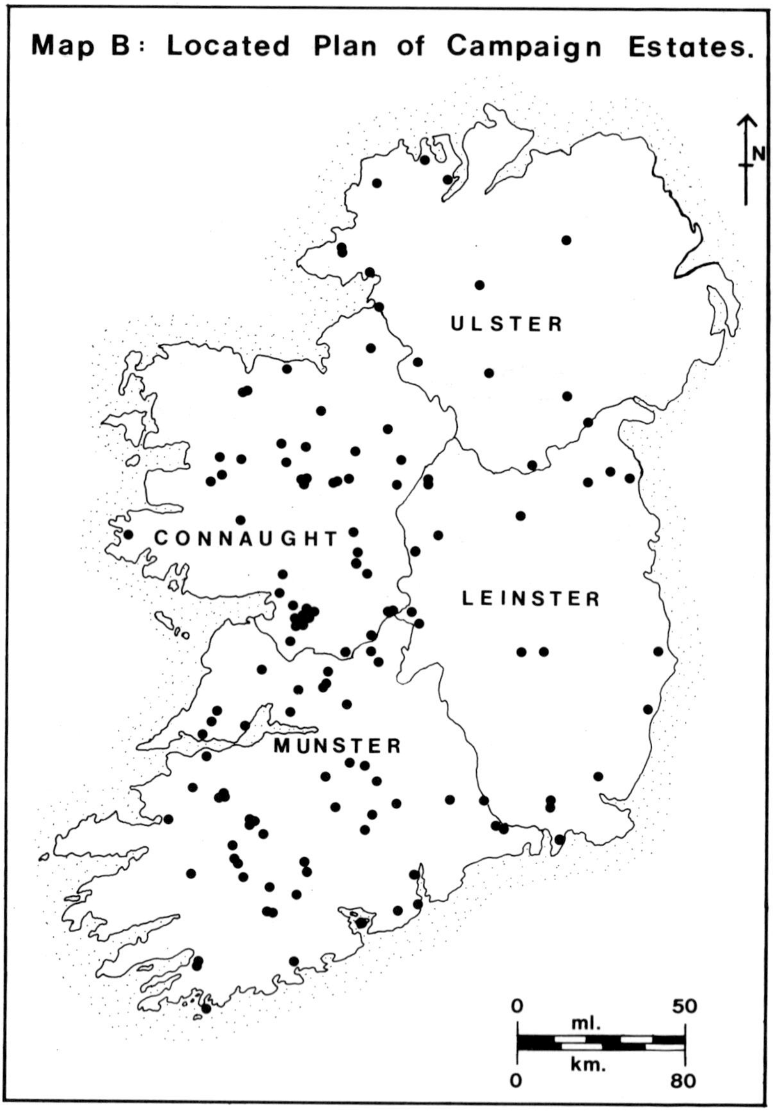
Map B: Located Plan of Campaign Estates.
N
ULSTER
CONNAUGHT
LEINSTER
MUNSTER
0
50
ml.
km.
0
80

APPENDIX 2

PLAN OF CAMPAIGN ESTATES

SYMBOLS:

* = information not available

( ) = source

? = unclear

SOURCES:
ACREAGE and VALUATION were compiled from PP1876 [C. 1492]. LXXX.35 *Return of the owners of land of one acre and upwards in the several counties, counties of cities and counties of towns in Ireland.*

U. H. Hussey De Burgh, *The landowners of Ireland. An alphabetical list of the owners of estates of 500 acres or £500 valuation and upwards in Ireland with the acreage and valuation in each county* (Dublin, 1878).

John Bateman, *The great landowners of Great Britain and Ireland* (New York, 1971).

NOTE:
Where two sets of figures are given, the first set represents the acreage and valuation directly affected by the Plan of Campaign; the second, in parantheses, indicates a landlord's complete holdings and their valuation.

	Estate	Acreage	Valuation	Plan Adopted	Tenants' Demands	Result
1	Alexander Co. Sligo	?	?	March 1888	Offer of 10% was considered inadequate (*UI*, 31.3.1888)	★
2	Lord Annesley Co. Cavan	24,222 (51,060)	8,802 (29,539)	December 1886	30% (*UI*, 18.12.1886)	Tenants agreed to buy holdings at 14 years' purchase on 11.8.1887 (*UI*, 13.8.1887)
3	Ashley Ballyhaunis, Co. Mayo	?	?	★	15% (*FJ*, 20.1.1887)	On 19.1.1887, landlord granted 15%; paid all law costs (*FJ*, 20.1.1887)
4	Atwell Co. Tyrone	822	177	Nov/Dec 1886	25% (*UI*, 18.12.1886)	Landlord granted 20% (*UI*, 19.3.1887)
5	Baggot Co. Galway	?	?	★	★	Settled June 1887 (TCD Ms. 6820 J. Dillon Papers, 318/Expenses of the Land War in 1889)
6	John Bagot Ballinturin House, Gort, Co. Galway; and Co. Roscommon	6,907 104	2,667 56	20 Jan 1887	Landlord offered 20% to non-judicial tenants but refused to reduce the judicial rents; tenants rejected these terms (*UI*, 22.1.1887)	Landlord granted 20% to non-judicial tenants and 15% to judicial; paid all law costs (*UI*, 11.6.1887)
7	Rev. H. L. Baker Co. Kilkenny	145	70	★	★	Landlord granted 40%; paid all law costs; forgave all arrears; tenants to buy their holdings at 16 years' purchase (*FJ*, 17.11.1890)

	Estate	Acreage	Valuation	Plan Adopted	Tenants' Demands	Result
8	Ball Co. Armagh	5,086	4,071	April 1887	30% (*FJ*, 9.4.1887)	On 17.8.1887, landlord granted 20% to judicial tenants and 25% to non-judicial; paid all law costs (*UI*, 3.9.1887)
9	Ban(n)on Broughal, Birr, Kings County	3,896	1,211	January 1887	Tenants objected to the appointment of Mr Garvey as land agent. (SPO, CSORP 1887/2222)	Tenants abandoned Plan in December 1889. (PRO Co 903/1, Irish Crime Records, 1885–1892, Intelligence Notes IV)
10	Barrett Carriganass, Bantry, Co. Cork	818	333	November 1886	25% (*UI*, 27.11.1886)	*
11	Barry Co. Sligo	?	?	*	'a reasonable reduction' (*FJ*, 22.1.1887)	*
12	A. H. Smith-Barry Tipperary	8,621 (26,792)	12,131 (37,279)	Mid-1889	withdrawal of Smith-Barry from Ponsonby syndicate and, subsequently, 25% rent reduction.	Settlements began in February 1891; completed in 1895. (Marnane, 'Humphreys' in Nolan (ed), *Tipperary*, pp.375–7)
13	Bell Co.Galway	?	?	*	*	Settled 1888. Tenants owed 2 years' rent; landlord settled for 1, less 17%, and paid all law costs (NLI Ms.13,502 (17), T. P. Gill Papers)
14	Jackson Bennett Cobh, Co. Cork	1,851	859	12 Dec. 1886	40% (*FJ*, 15.12.1886)	Landlord surrendered on tenants' terms immediately (*FJ*, 15.12.1886)

	Estate	Acreage	Valuation	Plan Adopted	Tenants' Demands	Result
15	Dean Massy-Beresford Lisnaskea, Co. Fermanagh	1,169 (7,793)	670 (5,481)	December 1886	★ (*UI*, 18.12.1886)	★
16	Berridge Clifden Castle, Connemara	159,898 (170,517)	6,321 (9,503)	★	★	Settled 1887 (TCD Ms.6820 Dillon/318)
17	Beytagh Mannin, Ballyhaunis, Co. Mayo	996	313	★	25% (NLI Ms.13,502 (17), T. P. Gill Papers)	Settled 1888. Landlord granted 20%, paid all costs (Gill Papers)
18	Mrs Blake Meelick, Whitegate, Co. Galway	?	?	★	★	Landlord granted 17½%, paid all costs (*UI*, 10.9.1887)
19	Charles Blake Co. Galway	?	?	13 Dec. 1886	35% (*FJ*, 15.12.1886)	★
20	A. Blennerhassett Ballyseedy, Tralee Co. Kerry	12,621	4,157	★	20%	Landlord yielded on tenants' terms, paid all law costs (*UI*, 12.5.1888)
21	A. Blennerhassett Knocknagree, Co. Cork	2,375	630	1887	★	In October 1890, tenants paid one year's rent and were forgiven that of two (*CE*, 14.10.1890)
22	Mrs Bodkin Cloondoreen, Co. Galway	?	?	★	★	Settled by an 'agreement satisfactory to the tenants and acceptable to Mrs Bodkin' (*FJ*, 12.9.1888)
23	Mrs Bourine Co. Cork	?	?	★	★ (*UI*, 20.11.1887)	★
24	Boxwell Co. Wexford	?	?	★	★ (*FJ*, 31.1.1887)	★

	Estate	Acreage	Valuation	Plan Adopted	Tenants' Demands	Result
25	Brabazon Cloonean, near Castlerea, Co. Roscommon	1,203	172	*	40% (*UI*, 3.12.1887)	*
26	Miss Bracken Blacklion, Co. Cavan	51 (1,319)	30 (402)	November 1886	25%	Landlord yielded on tenants' terms, paid all costs (*FJ*, 25.5.1887)
27	George F. Brooke Coolgreany, Co. Wexford	5,798	2,677	15 Dec. 1886	30% (*FJ*, 16.12.1886)	The estate was sold to the Land Commission and evicted tenants were restored between 1905–1907 (HC 1907 (C.3570) LXX 1155)
28	Browne Timicat, Co. Galway	?	?	8 Jan. 1888	40% (SPO, CSORP, 1888/7644)	*
29	Browne Cahermurphy, Co. Clare	?	?	*	*	Landlord granted 45%, forgave all arrears (*FJ*, 4.10.1888)
30	Bellew-Bryan Grangegeeth, Co. Meath	?	?	November 1886	20% (NLI Ms.13502 (3), Gill Papers)	On 2 April 1887, landlord granted 20% on rents fixed before 1884 and 15% on rents fixed since, paid all law costs (*UI*, 9.4.1887)
31	A. H. Burdett Banagher, Kings County	2,732 (3,965)	1,332 (2,155)	13 Dec. 1886	20% (*FJ*, 15.12.1886)	Landlord yielded on tenants' terms, paid all law costs (*UI*, 22.10.1887)

	Estate	Acreage	Valuation	Plan Adopted	Tenants' Demands	Result
32	D. F. Burke Co. Galway	?	?	November 1886	*	Landlord granted 20% to judicial tenants and 15% to non-judicial ones, paid all law costs (*FJ*, 15.12.1886)
33	Sir H. Burke Marble Hill, Loughrea, Co. Galway	25,259 (27,488)	7,564 (8,339)	Nov./Dec. 1886	30% (NLI Ms.13502 (18), Gill Papers)	At end of February 1888, landlord granted 25%, paid all law costs, paid £160 to support tenants he had evicted, re-instated the evicted, paid £140 compensation for crops (Gill Papers)
34	Mrs Burnaby Creewood, near Drogheda, Co. Louth	?	?	November 1886	20% (NLI Ms.13502 (5), Gill Papers)	Settled 4.1.1887 (TCD Ms.6820 Dillon/318)
35	Burton Co. Clare	?	?	January 1887	30% on judicial, 50% on non-judicial rents (FJ, 21.2.1888)	Landlord granted 30% to judicial and 50% to non-judicial tenants (*FJ*, 21.2.1888)
36	Butler Co. Clare	?	?	October 1887	20% on judicial rents (*UI*, 8.10.1887)	*
37	Hon. Mrs Butler Co. Kilkenny	?	?	November 1886	30% (*UI*, 27.11.1886)	*
38	J. E. Byrne Coolroe, Kinnagh, New Ross, Co. Wexford	535	309	*	*	Estate sold to tenants in 1907 (HC 1907 (C.3570) LXX.1155)

	Estate	Acreage	Valuation	Plan Adopted	Tenants' Demands	Result
39	Mrs Carroll Kilfinane, Co. Limerick	?	?	*	15% (*FJ*, 11.3.1887)	Landlord granted 15%, paid all law costs (*FJ*, 11.3.1887)
40	Caulfield Co. Roscommon	?	?	13 Jan. 1887	25% (*FJ*, 15.1.1887)	Landlord granted 20%, paid all law costs (*FJ*, 2.3.1887)
41	W. Chapman Co. Westmeath	?	?	February 1887	25% (*UI*, 19.2.1887)	Landlord granted 15%, paid all law costs (*UI*, 28.5.1887)
42	Cheever Co. Sligo	?	?	November 1887	50% (*UI*, 26.11.1887)	Settled 1888 (TCD Ms.6820 Dillon/318)
43	Lord Clancarty Garbally, Ballinasloe, Co. Galway	23,897 (25,510)	11,725 (12,817)	December 1886	25% (NLI Ms.13502 (16) Gill Papers)	On 23.6.1887 Landlord granted 20% (*FJ*, 24.6.1887)
44	Lord Clanricarde Portumna Castle, Co. Galway	49,026 (52,601)	19,635 (20,836)	16 Nov. 1886	40% (*FJ*, 17.11.1886)	Not finally settled until 1913 when the estate was sold to the tenants (Finnegan, 'Land war in s. e. Galway', p.133)
45	Clarke Borrisokane, Co. Tipperary	5,363 (6,122)	3,061 (3,168)	*	25% (SPO, CSORP 1891/24048)	Landlord accepted 4 years' rent, waived 2 years', allowed time for payment of a further one. (*FJ*, 23.9.1891)
46	Lord Clonbrock Ahascragh, Co. Galway	28,247	11,442	8 Dec. 1886	25% (*FJ*, 9.12.1886)	In May 1887 landlord granted 15% (SPO, P & CR, CBS, 1887-1898, Nos 1-3)

	Estate	Acreage	Valuation	Plan Adopted	Tenants' Demands	Result
47	S. D. Collis nr. Castleisland, Co. Kerry	3,598	1,207	January 1887	★ (*UI*, 15.1.1887)	★
48	I. D. C. Coltsman nr. Knocknagree, Co. Cork	4,478 (14,794)	1,309 (2,095)	★	★ (SPO, CSORP 1891/28048)	★
49	Marquis of Conyngham Mount Charles, Glenties, Co. Donegal	122,300 (166,710)	15,167 (50,076)	Nov./Dec. 1887	'a reasonable reduction' (*FJ*, 27.11.1888)	Landlord 'granted the reduction', paid all law costs, re-instated the evicted (*FJ*, 27.11.1888)
50	M. Cormack Modeshill, Co. Tipperary	?	?	29 May 1887	30% (*FJ*, 10.5.1888)	Settled June 1890 (SPO, P & CR, CBS, 1887-1898, No.5)
51	Captain Cosby Stradbally, Queens County	10,110	6,738	3 Dec. 1886	20% (*FJ*, 4.12.1886)	★
52	Costelloe Kilfree, Gurteen, Co. Sligo	1,330 (1,504)	413 (541)	March 1887	20% (*FJ*, 8.3.1887)	Settled 1887 (TCD Ms.6820 Dillon/318)
53	Captain P. Cowan Prospect, Eyrecourt, Co. Galway	2,373	1,008	December 1886	★ (*FJ*, 13.12.1886)	★
54	J. W. B. Creagh Donoughmore, Co. Cork	?	?	January 1887	30% (*FJ*, 13.1.1887)	★
55	Crofton Minors Co. Cork	?	?	February 1887	30% (*CE*, 15.3.1887)	In March 1887 landlord granted 20% (*CE*, 15.3.1887)

	Estate	Acreage	Valuation	Plan Adopted	Tenants' Demands	Result
56	P. J. B. Daly Ballinrobe, Co. Mayo	171	88	December 1886	20% (*FJ*, 28.2 / 16.3.1887)	Settled 1888 (TCD Ms.6820 Dillon/318)
57	Captain Dawson Tang, Co. Westmeath	?	?	November 1886	20% on judicial, 30% on non-judicial rents (*FJ*, 6.12.1886)	On 4.12.1886 landlord yielded on tenants terms (*FJ*, 6.12.1886)
58	Lord De Freyne (1) Frenchpark, Co. Roscommon	25,437 (29,817)	10,924 (12,571)	2 December 1886	20% (*FJ*, 3.12.1886)	In mid-January 1888 landlord allowed 25% on year's rent due to May 1886, 30% on rent due to May 1887; re-instated 56 evicted tenants, paid all law costs (*FJ*, 23/27.1.1888, 9.4.1888)
	(2)			Autumn 1888	★	Settled end January 1889, 30% offered on rental under £50, 15% on rental over £50 (SPO, P & CR, 1887-1917, Western Div. January 1889)
	(3)			Autumn 1889	30% (PRO, Co 903/4)	In December 1889 tenants accepted 15% (PRO, Co 903/4)
	(4)			November 1890	★	Landlord offered 30% to non-judicial, zero to judicial tenants (*The Times*, 6.2.1891)
59	John C. Delmege Glensharrold, Co. Limerick	3,067 (4,784)	1,173 (1,719)	6 Nov. 1887	57% (PRO, Co 903/1 No.XIII)	In April 1891 tenants accepted a reduction of 30% (PRO, Co 903/1)

	Estate	Acreage	Valuation	Plan Adopted	Tenants' Demands	Result
60	D'Estere Rossmanaher, Sixmilebridge, Co. Clare	2,833 (3,147)	1,625 (2,108)	★	★	Settled 1887 (TCD Ms.6820 Dillon/318)
61	Lord Dillon Co. Mayo	83,749 (94,764)	19,232 (28,762)	26 Nov. 1886	25% (*FJ*, 27.11.1886)	Landlord granted 20%, re-instated the evicted, paid all costs (*FJ*, 10.1.1887)
62	John P. Dillon R.M. Castlerea, Co. Roscommon	?	?	December 1887	★ (*UI*, 31.12.1887)	★
63	A. Dolphin Turoe, Loughrea, Co. Galway	1,171	630	13 Dec. 1886	40% (*FJ*, 15.12.1886)	★
64	Domville Balla, Co. Mayo	6,041 (9,066)	2,056 (4,876)	3 Dec. 1886	30% (*FJ*, 11.12.1886)	Settled 16.3.1887 (TCD Ms.6820 Dillon/318)
65	Dunne Athy, Co. Kildare	?	?	★	★ (*UI*, 11.12.1886)	★
66	Lord Dunsandle (1) Loughrea, Co. Galway	33,543 (37,057)	11,861 (17,393)	29 Nov. 1886	25% (*FJ*, 1.12.1886)	On 12.2.1887 landlord granted 20%, re-instated the evicted, paid all costs (*FJ*, 14.2.1887)
	(2)			December 1887	45% (*FJ*, 15.12.1887)	Landlord granted 35%, paid all costs (*FJ*, 23.4.1888)
67	Egan Meelin, Co. Cork	?	?	★	30% (*UI*, 27.8.1887)	On 19.8.1887 landlord granted tenants' demand (*UI*, 27.8.1887)

	Estate	Acreage	Valuation	Plan Adopted	Tenants' Demands	Result
68	Earl of Egmont Lohort Castle, Mallow, Co. Cork	16,766 (34,972)	13,594 (35,510)	7 Jan. 1887	30% (*FJ*, 10.1.1887)	★
69	Ellard Oola, Co. Limerick	866 (1,109)	1,024 (1,134)	January 1887	20% (*FJ*, 22.1.1887)	In March 1887 landlord granted 20%, paid all costs (*UI*, 2.4.1887)
70	Rev Conyngham Ellis Abbeyfeale, Co. Limerick	1,550	536	December 1887	50% (*UI*, 16.2.1889)	Tenants agreed to pay one year's rent less 40%, all arrears were cancelled, evicted tenants reinstated, landlord agreed to pay all costs (*UI*, 16.2.1889)
71	Marquis of Ely Loftus Hall, Fethard, Co. Wexford	14,023 (49,362)	9,168 (23,432)	★	★	Landlord granted 50% (*UI*, 17.12.1887)
72	Eyre Co. Galway	?	?	January 1887	25% (SPO, CSORP 1887/1519)	★
73	T. J. Eyre Figlash, nr Carrick-on-Suir, Co. Tipperary	763 (5,284)	556 (4,809)	September 1888	40% (SPO, CSORP 1891/24048)	Ended 22.6.1891 (*Irish Times*, 23.6.1891)
74	John Nolan Ferrell Logboy House, Ballyhaunis, Co. Mayo	9,732 (10,600)	3,072 (3,446)	End of 1886	20% (*FJ*, 28.2.1887)	Landlord granted 20%, paid all costs (*UI*, 6.8.1887)
75	Mrs Fetherstonhaugh Sperin Lodge, Co. Tyrone	?	?	November 1886	30% (*UI*, 27.11.1886)	★

	Estate	Acreage	Valuation	Plan Adopted	Tenants' Demands	Result
76	Fitzgerald Turlough Park, Castlebar, Co. Mayo	8,359 (8,367)	3,816 (3,825)	★	★	Settled 1887 (TCD Ms.6820 Dillon/318)
77	W. Fitzgerald Corofin, Co. Clare	?	?	10 Dec. 1886	30% (*FJ*, 13.12.1886)	Landlord granted 40%, paid all costs, cancelled arrears, reinstated the evicted (*FJ*, 5.1.1889)
78	Judge Flanagan Co. Clare	869 (4,442)	214 (2,049)	8 Jan. 1887	30% (*FJ*, 11.1.1887)	★
79	Flatley Co. Mayo	?	?	★	★	Settled 1889 (TCD Ms.6820 Dillon/318)
80	Colonel Fosberry nr Listowel, Co. Kerry	1,438	414	14 Dec. 1886	30% (SPO, CSORP 1886/22643)	Settled 1888 (TCD Ms.6820 Dillon/318)
81	Foy Co. Tipperary	?	?	February 1887	35% (*UI*, 12.2.1887)	On 29.9.1887 landlord granted 25% to judicial, 35% to non-judicial tenants, paid all costs (*UI*, 15.10.1887)
82	G. Furnell Co. Carlow	?	?	December 1886	20% (*FJ*, 7.12.1886)	★
83	Gardiner Caherhurley, Bodyke, Co. Clare	?	?	November 1886	20% (*FJ*, 3.3.1887)	Landlord granted 20% (*FJ*, 3.3.1887)
84	Gavins Shanchool, Co. Clare	?	?	27 Jan. 1887	Rents at the valuation (*UI*, 29.1.1887)	★
85	Mrs Gibbons Kilmovee, Co. Mayo	?	?	16 Nov. 1887	40% (*FJ*, 31.12.1887)	Landlord granted 30%, paid all costs (*UI*, 9.3.1889)
86	D. J. Gilliland Co. Donegal	3,940 (4,022)	1,249 (1,345)	March 1887	15% (*UI*, 19.3.1887)	★

	Estate	Acreage	Valuation	Plan Adopted	Tenants' Demands	Result
87	W. M. Glascott Alderton, New Ross, Co. Wexford	2,821 (3,290)	1,702 (2,341)	*	25% (*UI*, 12.2.1887)	Landlord granted 25%, each side paid own costs (*UI*, 12.2.1887)
88	James Gollock Muslera, Co. Cork	2,346	587	*	Offer of 20% was considered insufficient (*UI*, 22.1.1887)	*
89	Goold Abbeyfeale, Co. Limerick	10,966	3,090	26 Dec. 1886	* (*FJ*, 28.12.1886)	*
90	Sir Charles Gore Belleek Manor, Ballina, Co. Mayo	22,023 (30,592)	8,294 (11,082)	*	*	Landlord granted 15%, reinstated evicted, paid costs (*FJ*, 10.5.1887)
91	Earl of Granard Castleforbes, Newtownforbes, Co. Longford	14,979 (21,294)	6,637 (9,840)	December 1886	30% judicial, 40% non-judicial (*FJ*, 9.12.1886)	*
92	E. Greer Co. Sligo	?	?	March 1887	20% (*UI*, 12.3.1887)	*
93	T. Gregory Carrickmoira, Newtownforbes, Co. Longford	625	273	December 1887	25% (*UI*, 17.12.1887)	*
94	Griffith Co. Clare	4,949	1,132	*	*	Landlord granted 20% on rents and arrears, paid all costs (*UI*, 4.6.1887)
95	Mrs Hagerty Borris, Co. Carlow	1,256	792	December 1886	15% (*UI*, 11.12.1886)	*

	Estate	Acreage	Valuation	Plan Adopted	Tenants' Demands	Result
96	Captain Hall Co. Galway	?	?	★	★	Settled 1888 (TCD Ms.6820 Dillon/318)
97	Hamilton Co. Armagh	?	?	April 1887	★ (*FJ*, 9.4.1887)	★
98	Rev J. Hamilton Cullmore House, Ballyshannon, Co. Donegal	?	?	End of 1886	15% (*FJ*, 11.11.1887)	Landlord granted 15% (*FJ*, 11.11.1887)
99	Mrs M. M Hamilton nr Mallow, Co. Cork	?	?	★	★ (*CE*, 29.11.1887)	★
100	Lord Harlech Co. Mayo	2,546 (58,358)	606 (26,400)	16 Dec. 1886	25% (*FJ*, 17.12.1886)	Settled 1889 (TCD Ms.6820 Dillon/318)
101	Hayes Co. Limerick	2,057	530	★	★	Settled 1887 (TCD Ms.6820 Dillon/318)
102	Lord Headfort Virginia, Co. Cavan	14,221 (42,754)	9,013 (39,606)	November 1886	★ (*UI*, 20.11.1886)	★
103	Major Hickman Newmarket-on-Fergus Co. Clare	3,030	1,419	December 1886	★ (*FJ*, 17.10.1887)	★
104	Captain Hill Gweedore, Co. Donegal	?	?	December 1886	33⅓% (*FJ*, 11.12.1886)	Landlord granted 30%, reinstated 130 evicted tenants, paid £900 costs, granted a 60% abatement on arrears, which reduced the amount owed by the tenants from £3,400 to £1,450 (*FJ*, 25.11.1887)

	Estate	Acreage	Valuation	Plan Adopted	Tenants' Demands	Result
105	Hillas Dromore, Co. Sligo	7,445 (7,771)	2,352 (2,714)	13 Mar. 1887	at least 20% (*UI*, 19.3.1887)	*
106	S. N. Hutchins J. P. Ardnagoshel, Bantry Co. Cork	7,859 (8,220)	3,223 (3,498)	January 1887	25% (*CE*, 12.1.1887)	Landlord granted 15% (*UI*, 25.6.1887)
107	J. Johnston J. P. Ballsmill, Co. Louth	?	?	February 1887	* (*UI*, 12.2.1887)	*
108	Lord Kenmare Killarney, Co. Kerry	91,080 (118,606)	25,253 (34,473)	28 Oct. 1888	20% on judicial, 40% on non-judicial rents (*FJ*, 29.10/29.12.1888)	25% reduction given to yearly tenants, 20% to leaseholders and judicial tenants, evicted tenants reinstated. Arrears and legal costs were wiped out on payment of one year's rent. (NLI Ms.13500 (6), Gill Papers)
109	Kennedy Kilmore, Co. Roscommon	503	273	*	*	Settled 1889 (TCD Ms.6820 Dillon/318)
110	Kenny Enniskeen, Co. Louth	90 (1,926)	61 (1,633)	*	*	30% reduction given to judicial, 50% to non-judicial tenants (*UI*, 20.8.1887)
111	Lady Kingston Mitchelstown Castle, Co. Cork	24,671	17,951	11 Dec. 1886	20% (*CE*, 13.12.1886)	20% granted, evicted reinstated, landlord paid costs (*CE*, 13.2.1888)

	Estate	Acreage	Valuation	Plan Adopted	Tenants' Demands	Result
112	Lord Kingston Kilronan Castle, Keadue, Carrick-on-Shannon, Co. Roscommon	811	528	November 1886	20% (*FJ*, 27.11.1886)	*
113	Mrs Kirwan Co. Galway	?	?	24 Jan. 1887	25% (*FJ*, 26.1.1887 *UI*, 29.1.1887)	Settled 1887 (TCD Ms.6820 Dillon/318)
114	Knox Ballina, Co. Mayo	?	?	December 1886	25% (*FJ*, 7.12.1886)	Tenants accepted 20% (*UI*, 21.5.1887)
115	Arthur Langford Rowls, Newmarket, Co. Cork	2,440	393	September 1887	20% (*CE*, 13.11.1888)	*
116	Lord Lansdowne Luggacurran, Queens County	8,981 (142,916)	5,310 (62,025	17 Nov. 1886	20% on judicial, 25 to 35% on non-judicial rents (*Kildare Observer*, 20.11.1886)	Estate sold to tenants between 1905-1907 (HC 1907 (c.3570) LXX.1155)
117	W. N. Leader Curass and Meelin, Co. Cork	?	?	18 Dec. 1886	20% on judicial, 25% on non-judicial rents NLI Ms.13502 (12) Gill Papers	Paid 2 years' rent in lieu of all arrears up to September 1890, agreed to buy farms at 14 years purchase (*FJ*, 5.5.1891)
118	Leetch Kilkelly, Co. Mayo	329	75	*	*	Settled 1888 (TCD Ms.6820 Dillon/318)
119	Mrs Hannah Lewis Ballinagar, Loughrea, Co. Galway	6,684	1,339	13 Dec. 1886	50% (*FJ*, 15.12.1886)	In 1908 estate sold to tenants (Land Commission ETA/6777)
120	Lewis Inniskeen, Co. Monaghan	2,489 (3,773)	1,608 (2,863)	*	*	Settled 1888 (TCD Ms.6820 Dillon/318)

	Estate	Acreage	Valuation	Plan Adopted	Tenants' Demands	Result
121	W. Lloyd Tower Hill, Cappamore, Co. Limerick	968	870	December 1886	30% (*FJ*, 17.12.1886)	Settled April 1891. Tenants bought farms at 20 years' purchase of government valuation (*CC*, 14.4.1891)
122	London Companies, Drapers & Skinners Co. Derry	?	?	3 Dec. 1886	* (*FJ*, 4.12.1886, SPO, CSORP 1886/22318)	Dispute settled by sale of farms to tenants (*ETC*, 2797-2812)
123	W. Lysaght Meelin, Co. Cork	2,252	382	December 1886	30% (SPO, CSORP 1887/2189)	*
124	J. McDermot Rathmore, Co. Galway	?	?	December 1886	* (*FJ*, 13.12.1886)	*
125	McGwyre or Maguire Tynagh, Loughrea, Co. Galway	?	?	4 Feb. 1887	25% (NLI Ms.13502 (9), Gill Papers)	Settled April 1887 (Gill Papers)
126	Sir J. N. McKenna MP Killeagh, Co. Cork	1,582 (2,934)	1,173 (504)	13 Dec. 1886	*	Landlord granted 20% (*UI*, 5.3.1887)
127	W. McMahon Mountfield, Omagh, Co. Tyrone	16,326 (21,029)	3,375 (5,302)	Feb./Mar. 1887	20% (*UI*, 12.3.1887)	Landlord granted 20% (*UI*, 12.3.1887)
128	R. Macarick Enniscrone, Swinford, Co. Mayo	3,453	685	January 1887	* (*FJ*, 13.1.1887)	Landlord granted 30%, paid all costs (*UI*, 18.2.1888)
129	Sir James Mackey Knocknagree, Co. Cork	1,377	243	January 1887	30% on judicial, 50% on non-judicial rents (*CE*, 15.1.1887)	On 7.10.1889 tenants agreed to pay 1 year's rent, less reduction of 50% and were given clear receipt to May 1889 (*CE*, 9.10.1889)

	Estate	Acreage	Valuation	Plan Adopted	Tenants' Demands	Result
130	J. Maconchy Evaghan, Co. Longford	?	?	December 1887	40% (*UI*, 24.12.1887)	★
131	Packenham-Mahon Strokestown, Co. Roscommon	26,980 (28,123)	14,116 (15,080)	★	40% (*UI*, 26.11.1887)	In September 1888 landlord granted 15% to judicial, 20% to non-judicial tenants
132	Sir W. Ross Mahon Castlegar, Ahascragh, Co. Galway	8,619 (9,479)	3,788 (4,358)	January 1887	30% (*FJ*, 28.2.1888)	Landlord granted 45%, paid all costs, cancelled all arrears (*FJ*, 28.2.1888)
133	T. R. Marmion Islands off west Cork coast	?	?	July 1890	★ (*CE*, 8.7.1890)	★
134	Lord Massereene Collon, Co. Louth	7,193 (21,024)	5,234 (15,013)	February 1887	20% on judicial, 25% on non-judicial rents (ETC, 6142-7)	Estate sold to tenants between 1905 - 1907 (HC 1907 (C.3570) LXX 1155)
135	Mrs E. Massey Macroom, Co. Cork	13,363	3,106	13 Jan. 1889	★ (*CE*, 14.1.1889)	'a settlement satisfactory to the tenants' (*CE*, 30.4.1891)
136	G. W. Massey Co. Tipperary	?	?	December 1888	20% (*FJ*, 13.12.1889)	On 12.12.1889 landlord granted 22½% (*FJ*, 13.12.1889)
137	Medlycott Rocket Castle, Portlaw, Co. Waterford	958	1,348	★	★	Settled 1887 (TCD Ms.6820 Dillon/318)
138	Meldon Coolarne, Turloughmore, Co. Galway	4,867 (5,864)	2,571 (1,825)	★	★	Settled 1888 (TCD Ms.6820 Dillon/318)

	Estate	Acreage	Valuation	Plan Adopted	Tenants' Demands	Result
139	Mrs Moreland Tomgraney, Co. Clare	5,118 (5,793)	1,653 (2,054)	10 Dec. 1886	25% (*FJ*, 12.12.1886)	Settled 1887 (TCD Ms.6820 Dillon/318)
140	John Murphy Co. Roscommon	5,363 (6,515)	2,039 (2,477)	December 1886	*	Settled 1888 (*FJ*, 30.6.1888)
141	Mrs Naghten nr Athlone, Co. Westmeath	?	?	8 Dec. 1886	25% (*FJ*, 14.12.1886)	On 26.5.1887 landlord granted 20%, paid costs, reinstated the evicted (*UI*, 11.6.1887)
142	Harper C. Nixon Co. Donegal	9,084 (9,330)	554 (728)	*	* (SPO, CSORP 1891/24048)	*
143	Lady Nugent Ballinlough Castle, Delvin, Co. Westmeath	4,692 (7,008)	3,720 (4,598)	28 Nov. 1886	20% (*FJ*, 29.11.1886)	Landlord granted 15% (*UI*, 23.4.1887)
144	P. O'Brien Durr, Co. Limerick	?	?	Nov/Dec. 1886	20% (*FJ*, 2.12.1886)	Settled 1888 (TCD Ms.6820 Dillon/318)
145	Captain C. G. O'Callaghan Ballinahinch, Tulla, Co. Clare	8,271	3,590	December 1886	25% (*FJ*, 17.12.1886)	Landlord granted 20%, cancelled 1 year's arrears (*UI*, 29.1.1887)
146	Colonel John O'Callaghan Bodyke, Co. Clare	4,842	1,920	30 Jan. 1887	*	£1,000 paid to landlord. Gave clear receipt to 72 tenants to March 1887, paid all costs, re-instated the evicted (*FJ*, 4.2.1888)
147	D. O'Connell Kilgory, Tulla, Co. Clare	2,019	946	December 1886	20% below Griffith's Valuation (*FJ*, 28.12.1886)	Landlord granted from 12½% to 25%, paid all costs (*UI*, 26.5.1888)

	Estate	Acreage	Valuation	Plan Adopted	Tenants' Demands	Result
148	P. O'Gara Co. Mayo	?	?	November 1886	⋆	Landlord granted 'Lord Dillon's terms' (*FJ*, 9.3.1887)
149	The O'Grady Kilballyowen, Bruff, Co. Limerick	1,943	2,232	9 Nov. 1886	30% on judicial, 40% on non-judicial rents (*FJ*, 10.11.1886)	⋆
150	Miss A. O'Hara Kilbeheny, Co. Limerick	255	177	October 1888	30% (*FJ*, 5.10.1888)	Landlady granted 30%, paid costs, cancelled arrears (*CE*, 16.2.1889)
151	Peter de Penthony O'Kelly (1) Clongorey, Co. Kildare	3,153	2,117	December 1886	20% on judicial 30% on non-judicial rents (*UI*, 1.1.1887)	In May 1887 tenants accepted 10% on judicial and 20% on non-judicial rents (*ETC*, 635-778)
	(2)			November 1888	⋆	⋆ (*ETC*, 635-778)
152	Wybrants Olphert Ballyconnell House, Falcarragh, Co. Donegal	18,133	1,802	January 1888	25% on judicial, 40% on non-judicial rents (PRO, Co 903/1, Irish Crime Records, XII)	⋆
153	Lord Ormathwaite Co. Kerry	8,907 (26,261)	2,620 (14,667)	January 1887	30% (*FJ*, 4.1.1887)	Settled 1888 (TCD Ms.6820 Dillon/318)
154	Anthony Ormsby Ballynamore, nr Balla, Co. Mayo	4,493	1,685	December 1886	30% (*FJ*, 24.5.1888)	Settled August 1888, tenants paid costs (SPO, P & CR, CBS, DICS, 1887-1898 Western Division)
155	H. B. O'Sullivan Carrigphooka, Macroom, Co. Cork	2,411	534	23 Feb. 1888	25% on judicial, 50% on non-judicial rents (*CE*, 25.2.1888)	⋆

	Estate	Acreage	Valuation	Plan Adopted	Tenants' Demands	Result
156	Perry Woodroffe, Clonmel, Co. Tipperary	2,769	2,873	November 1887	30% (*UI*, 12.11.1887)	*
157	Burton-Persse Moyode Castle, Craughwell, Co. Galway	6,469 (9,496)	2,920 (4,203)	*	*	Settled 1888 (TCD Ms.6820 Dillon/318)
158	Persse Roxborough, Loughrea, Co. Galway	12,394 (13,585)	4,998 (6,002)	January 1887	50% (*UI*, 5.2.1887)	*
159	Phibbs Co. Sligo	?	?	*	*	Settled 1888 (TCD Ms.6820 Dillon/318)
160	Pollock (1) Creggs, Co. Galway	?	?	December 1887	40% (*FJ*, 13.4.1888)	Landlord offered 30%, paid all costs (*FJ*, 13.4.1888)
	(2)			November 1888	40% (SPO, P & CR, 1887-1917, Western Div. November 1888 and April 1889)	On 29.4.1889 landlord granted 20% (SPO, P & CR, 1887-1917, April 1889)
161	C. W. T. Ponsonby Park House, Youghal, Co. Cork	10,367	6,768	14 Nov. 1886	25% on judicial, 35% on non-judicial rents (*FJ*, 15.11.1886)	Estate sold to tenants between 1905 - 1907 (HC 1907 (H.3570) LXX 1155)
162	Powell Co. Mayo	?	?	*	*	Settled 1888 (TCD Ms.6820 Dillon/318)
163	Quin Co. Longford	?	?	January 1887	* (SPO, CSORP 1887/1022)	Settled 1888 (TCD Ms.6820 Dillon/318)

	Estate	Acreage	Valuation	Plan Adopted	Tenants' Demands	Result
164	Captain Quinn Ballindunny, Co. Tipperary	?	?	13 Dec. 1886	20% (*FJ*, 15.12.1886)	20% granted (*UI*, 2.4.1887)
165	J. Rashleigh Co. Mayo	5,476 (36,896)	833 (11,291)	December 1886	★	On 18.1.1887 landlord granted 20% reduction, reinstated the evicted, paid all costs (*FJ*, 21.1.1887)
166	Richardson Co. Galway	?	?	★	★	Settled 1888 (TCD Ms.6820 Dillon/318)
167	Roe Co. Tipperary	?	?	January 1887	30% (*FJ*, 19.1.1887)	★
168	R. De Ros Rose Co. Carlow	?	?	December 1886	30% (*FJ*, 7.12.1886)	★
169	F. Russell Lissanode, Moate, Co. Westmeath	1,022 (1,514)	651 (925)	February 1887	20% (*UI*, 19.2.1887)	★
170	Rutagh Co. Limerick	?	?	November 1886	30% (*FJ*, 11.11.1886)	★
171	V. Ryan Ardara, Co. Donegal	5,163 (7,245)	407 (1,473)	15 Dec. 1886	20% on judicial, 33⅓% on non-judicial rents (SPO, CSORP 1887/10534)	Landlord granted 17½% on judicial rents, paid costs (*UI*, 3.12.1887)
172	Sampson Co. Clare	?	?	January 1887	25% (NLI Ms.13502 (17), Gill Papers)	In August 1888 landlord granted 25% reduction, paid all costs (Gill Papers)
173	T. G. Wills Sandford Castlerea, Co. Roscommon	24,411	10,194	January 1887	30% (*FJ*, 21.3.1887)	Settled July 1889 (SPO, P & CR, CBS, 1887-1889, Nos 1-3)

	Estate	Acreage	Valuation	Plan Adopted	Tenants' Demands	Result
174	Scott Kildysart, Co. Clare	6,431	2,953	December 1886	35% (*UI*, 24.12.1886)	Settled 1888 (TCD Ms.6820 Dillon/318)
175	Shirley Lough Fea, Farney, Co. Monaghan	26,386 (28,760)	20,744 (23,744)	25 Mar. 1887	* (*FJ*, 28.3.1887)	*
176	Sinnott Enniscorthy, Co. Wexford	984	754	*	*	Settled 1887 (TCD Ms.6820 Dillon/318)
177	Mrs Smith Cockermouth, Co. Monaghan	?	?	*	15% (*FJ*, 6.7.1888)	Landlord granted 15%, paid all costs (*FJ*, 6.7.1888)
178	Hon. Moore Smyth Ballinatray, Co. Waterford	7,125 (7,480)	3,610 (4,415)	Nov/Dec. 1886	* (*FJ*, 3.12.1886)	*
179	Stewart Rock Hill, Letterkenny, Co. Donegal	538 (5,089)	489 (1,544)	February 1887	33⅓ - 40% (*ETC*, 13984–995)	*
180	Murray Stewart Ardara, Co. Donegal	50,818 (98,269)	6,500 (22,822)	January 1888	33% on judicial, 40% on non-judicial rents (*UI*, 14.1.1888)	In March 1888 agent agreed to pay costs, tenants paid full judicial rents (NLI Ms.4275, Murray Stewart letterbook, p.498)
181	Hugh Swanzy Castleblaney, Co. Monaghan	501 (1,045)	399 (730)	*	25% (*UI*, 2.4.1887)	Landlord granted 25%, paid all costs (*UI*, 2.4.1887)
182	Swiney Moyagh, Ramelton, Co. Donegal	678	307	*	20% (*ETC*, 13,885–912)	*

	Estate	Acreage	Valuation	Plan Adopted	Tenants' Demands	Result
183	Taffe Co. Mayo	?	?	13 Dec. 1886	40 to 50% (*FJ*, 14.12.1886)	In August 1888 agent accepted 1 year's rent in lieu of arrears (*FJ*, 18.8.1888)
184	Tew Co. Tipperary	207	112	November 1886	20% (*UI*, 9.4.1887)	Landlord granted 20%, paid all costs (*UI*, 9.4.1887)
185	Colonel Thackwell Aghada, Co. Carlow	?	?	December 1886	30% (*The Times* 15.12.1886)	⋆
186	Colonel C. G. Tottenham Ballycurry, Ashford Co. Wicklow	2,540 (11,517)	1,409 (12,679)	⋆	⋆	⋆
187	Captain T. A. Townshend Ballinahulla, nr Kingwilliamstown, Co. Cork	?	?	⋆	⋆	On 22.5.1889 tenants bought farms at 17 years' purchase of the valuation (*CE*, 23.5.1889)
188	H. B. Trench Kylebeg and Gurtymadden, Co. Galway	1,581 (12,134)	798 (6,578)	⋆	30% (*UI*, 22.10.1887)	Landlord granted 25%, paid all costs (*UI*, 22.10.1887)
189	J. A. Trench Corlane, New Inn, Co. Tipperary	?	?	November 1886	25% (*UI*, 16.7.1887)	On 22.6.1887 agent offered 20% (*UI*, 16.7.1887)
190	W. T. Trench (1) Abbeyfeale, Co. Limerick	?	?	28 Dec. 1886	30% (*UI*, 13.8.1887)	Landlord granted 30%, paid all costs (*UI*, 13.8.1887)

	Estate	Acreage	Valuation	Plan Adopted	Tenants' Demands	Result
	(2)			December 1887	50% (NLI Ms.13502 (17), Gill Papers)	Settled 1888. Landlord granted 50%, paid costs, cancelled arrears. Tenants bought farms at 16 years' purchase (Gill Papers)
191	O'Hara Trench Clonfert House, Eyrecourt, Co. Galway	5,409	2,142	March 1887	33⅓% (NLI Ms.13502 (17), Gill Papers	In 1888 tenants accepted 27½% (Gill Papers)
192	Tuohy nr Kiltimagh, Co. Mayo	1,347	529	*	30% (*UI*, 18.2.1888)	On 10.2.1888 landlord granted 30% on judicial, 50% on non-judicial rents (*UI*, 18.2.1888)
193	Tuthill Castlegarde, Co. Limerick	1,107	1,252	*	*	On 9.4.1887 agent granted 20% on judicial, 30% on non-judicial rents (*FJ*, 13.4.1887)
194	Hastings Twiss Bird Hill, Co. Tipperary	2,706 (3,518)	1,748 (2,078)	*	15% on judicial, 25% on non-judicial rents (*UI*, 22.6.1889)	In February 1889 tenants agreed to pay 1 year's rent on 15 April and another on 11 June. They owed 3½ years (*UI*, 22.6.1889)
195	Col Hector S. Vandeleur Kilrush House, Co. Clare	19,790 (20,206)	11,216 (11,596)	15 Dec. 1886	25% on judicial, 40% on non-judicial rents (*FJ*, 16.12.1886)	Settled by arbitration on 4 May 1889 (*FJ*, 6.5.1889)
196	Mrs Vandeleur Corraclare, Co. Clare	?	?	March 1887	40% (*UI*, 26.3.1887)	Settled 1887 (TCD Ms.6820 Dillon/318)

	Estate	Acreage	Valuation	Plan Adopted	Tenants' Demands	Result
197	Waithman Co. Roscommon	4,631 (9,172)	1,911 (3,950)	*	*	Settled 1888 (TCD Ms.6820 Dillon/318)
198	Lord Wallscourt Ardfry, Oranmore, Co. Galway	2,828	1,970	December 1886	30% (*UI*, 30.6.1888)	Landlord granted 25%, paid all costs (*UI*, 30.6.1888)
199	Lord Waterford Curraghmore, Portlaw, Co. Waterford	39,884 (66,684)	27,705 (32,749)	November 1887	50% (*UI*, 5/12.11.1887)	Settled 1888 (TCD Ms.6820 Dillon/318)
200	Earl of Westmeath Tynagh, Loughrea, Co. Galway	14,604 (15,695)	4,378 (5,486)	November 1886	40% (*FJ*, 25.4.1888)	Landlord granted 30%, paid all costs (*FJ*, 25.4.1888)
201	Mrs F. R. Sadlier Wheeler nr Clonakilty, Co. Cork	?	?	March 1887	* (*CE*, 11.3.1887)	*
202	Worthington Co. Roscommon	2,573 (3,087)	405 (647)	November 1887	50% (*UI*, 26.11.1887)	*
203	Wynne Manorhamilton, Co. Leitrim	15,436 (28,418)	4,381 (14,091)	21 Dec. 1886	25% (*FJ*, 22.12.1886)	Settled 1887 (TCD Ms.6820 Dillon/318)

APPENDIX 3

PLAN OF CAMPAIGN ESTATES
BY COUNTY AND PROVINCE

CONNAUGHT

Galway	32
Leitrim	1
Mayo	20
Roscommon	11
Sligo	7
	71

MUNSTER

Clare	18
Cork	24
Kerry	5
Limerick	13
Tipperary	12
Waterford	3
	75

LEINSTER

Carlow	4
Dublin	–
Kildare	2
Kilkenny	2
Kings	2
Longford	4
Louth	4
Meath	1
Queens	2
Westmeath	5
Wexford	6
Wicklow	1
	33

ULSTER

Armagh	2
Antrim	–
Cavan	3
Derry	1
Donegal	10
Down	–
Fermanagh	1
Monaghan	4
Tyrone	3
	24

APPENDIX 4

PLAN OF CAMPAIGN
COMMENCEMENT AND SETTLEMENT DATES

A. Commencement Dates

The Plan of Campaign was adopted on 203 estates but, as the agitation was repeated on several properties, there were 210 instances in all. We know the commencement date in 157 cases, that is some 75% of the total.

1886

Nov	23
Nov/Dec	6
Dec	50
	79

1887

Jan	23
Feb	8
Feb/March	1
March	8
April	2
May	1
July	1
Sept	1
Oct	1
Nov	6
Nov/Dec	1
Dec	7
1887	1
	61

1888

Jan	3
Feb	1
March	1
Sept	1
Oct	2
Autumn	1
Nov	2
Dec	1
	12

1889

Jan	1
Mid-1889	1
Autumn	1
	3

1890

July	1
Nov	1
	2

TOTAL: 157

B. Settlement Dates

1886

Dec	3
	3

1887

Jan	5
Feb	2
March	10
April	7
May	7
June	6
Aug	6
Sept	2
Oct	2
Nov	2
Dec	2
1887	11
	62

1888

Jan	1
Feb	7
March	1
April	3
May	2
June	1
July	1
Aug	3
Sept	2
Oct	1
Nov	1
1888	22
	45

1889

Jan	2
Feb	3
March	1
April	1
May	2
July	1
Oct	1
Dec	3
1889	3
	17

1890

June	1
Oct	1
Nov	1
	3

1891

Feb	2
April	3
May	2
June	1
Sept	1
	9

1905–7	5
1908	1
1913	1
	7

TOTAL: 146

APPENDIX 5

IRISH AGRARIAN OUTRAGES, 1885–1891

SOURCE: SPO, Irish Crime Records, 1879–1893
Returns of Agrarian Outrages Reported to the Constabulary.

Offences Against the Person	**1885**	**1886**	**1887**	**1888**	**1889**	**1890**	**1891**
homicide: murder	6	7	6	6	1	2	2
homicide: manslaughter	1	3	2	1	–	4	–
firing at the person	12	16	19	14	11	15	8
assault with intent to murder	–	1	1	–	–	–	–
assault on police	7	4	11	1	3	1	–
aggravated assault	18	20	27	24	14	15	8
assault endangering life	8	11	9	5	6	9	2
assault on bailiffs and process servers	7	11	11	2	1	–	1
cutting or maiming the person	3	1	5	–	1	2	1

IRISH AGRARIAN OUTRAGES, 1885–1891

Offences Against Property	**1885**	**1886**	**1887**	**1888**	**1889**	**1890**	**1891**
incendiary fire and arson	94	103	125	73	74	5	77
burglary and housebreaking	3	–	2	5	1	2	–
robbery	7	21	5	16	14	3	6
taking and holding forcible possession	7	11	5	2	–	1	–
cattle stealing	29	26	17	10	16	7	5
sheep stealing	1	–	2	–	–	–	2
killing, cutting or maiming cattle	59	73	54	55	50	49	55
sacrilege	1	–	–	–	–	–	–

Offences Against the Public Peace	**1885**	**1886**	**1887**	**1888**	**1889**	**1890**	**1891**
demand or robbery of arms	9	2	9	1	–	3	–
riots and affrays	5	9	10	2	3	4	1
administering letters or notices	6	6	1	2	–	–	–
threatening letters or notices	438	424	292	249	194	199	215
intimidation otherwise	74	92	93	67	38	42	30
pound breach	1	6	1	–	1	3	2
attacking houses	1	1	2	–	1	1	2
resistance to legal process	2	10	14	10	8	2	1
injury to property	109	150	111	76	63	65	35
firing into dwellings	33	43	35	30	29	30	16
injury to place of worship	1	2	–	–	1	–	–
having arms or being armed in a proclaimed district	–	–	–	–	–	–	1

IRISH AGRARIAN OUTRAGES, 1885-1891

Other Offences	**1885**	**1886**	**1887**	**1888**	**1889**	**1890**	**1891**
injury to, or attempt to injure, railway trains or highways	2	3	9	7	4	1	1
injury to telegraph	–	–	4	1	–	–	–
perjury	–	–	1	1	–	–	1
TOTAL	944	1,056	883	660	534	519	472

By County and Province

ULSTER	**1885**	**1886**	**1887**	**1888**	**1889**	**1890**	**1891**
Antrim	3	2	1	1	–	1	2
Armagh	1	2	5	6	6	3	3
Cavan	19	14	4	4	11	17	9
Donegal	7	9	8	18	11	6	7
Down	6	5	4	4	5	6	6
Fermanagh	8	9	6	7	1	4	–
Londonderry	12	13	15	7	3	2	2
Monaghan	8	12	7	6	11	7	2
Tyrone	3	22	11	4	6	5	6
TOTAL	67	88	61	57	54	51	37

IRISH AGRARIAN OUTRAGES, 1885-1891

By County and Province

LEINSTER	**1885**	**1886**	**1887**	**1888**	**1889**	**1890**	**1891**
Carlow	6	2	1	2	–	–	5
Dublin	2	2	1	1	1	2	1
Kildare	7	4	4	3	4	2	5
Kilkenny	9	12	16	6	6	4	2
Kings	26	19	22	7	3	3	17
Longford	23	19	8	9	6	8	8
Louth	6	4	4	1	8	5	–
Meath	5	4	1	–	1	6	3
Queens	22	10	15	5	6	5	6
Westmeath	21	24	19	10	15	14	12
Wexford	25	12	13	17	9	7	3
Wicklow	4	2	4	2	2	–	2
TOTAL	156	114	108	63	61	56	64

CONNAUGHT	**1885**	**1886**	**1887**	**1888**	**1889**	**1890**	**1891**
Galway ER	36	72	78	61	40	48	20
WR	43	44	32	28	24	28	16
Leitrim	23	29	30	31	18	10	10
Mayo	26	31	45	22	29	24	8
Roscommon	14	31	16	13	10	9	11
Sligo	11	16	20	14	4	3	6
TOTAL	153	223	221	169	125	122	71

IRISH AGRARIAN OUTRAGES, 1885–1891

By County and Province

MUNSTER	**1885**	**1886**	**1887**	**1888**	**1889**	**1890**	**1891**
Clare	88	141	153	104	71	60	94
Cork ER	75	73	66	40	43	37	28
WR	45	44	33	34	39	36	19
Kerry	180	209	108	96	63	55	68
Limerick	61	67	56	34	25	26	17
Tipperary NR	38	23	34	24	8	13	16
SR	56	54	37	29	34	56	49
Waterford	25	20	6	10	11	7	9
TOTAL	568	631	493	371	294	290	300
GENERAL TOTAL	944	1,056	883	660	534	519	472

TOTAL OF OUTRAGES IN IRELAND

SOURCE: PRO, CO 903, RIC Statistics of Crimes Specially Reported to the Constabulary Office, quoted in Charles Townshend, *Political violence in Ireland: government and resistance since 1848* (Oxford, 1983), p.151.

Year	**Agrarian**	**Total**	**Percentage Agrarian**
1879	863	3,500	25
1880	2,585	5,669	46
1881	4,439	7,778	57
1882	3,433	6,628	51
1883	870	2,535	34
1884	762	2,453	31
1885	944	2,683	35
1886	1,056	3,251	35
1887	883	2,270	39
1888	660	2,182	30
1889	534	1,895	28
1890	519	1,934	27

APPENDIX 6

IRISH AGRICULTURAL STATISTICS, 1876 - 1891

A. CROP PRODUCTION

SOURCE: HC 1886 [c. 4802]. *The Agricultural Statistics of Ireland for 1885,* p.12, table VIII; HC 1893 [c. 7187]. *The Agricultural Statistics of Ireland for 1892,* p.17, table VIII.

	Wheat cwts	**Oats cwts**	**Barley cwts**	**Potatoes tons**	**Turnips tons**	**Hay tons**
1876	2,023,493	21,415,196	3,879,528	4,154,785	4,540,818	3,458,239
1877	1,901,100	17,846,563	3,507,764	1,757,275	3,564,025	4,311,163
1878	2,307,685	19,044,605	3,917,704	2,526,504	4,686,226	4,417,344
1879	1,798,931	15,532,629	3,260,810	1,113,676	2,057,804	3,598,580
1880	2,227,717	19,558,470	3,444,442	2,985,859	4,339,688	3,795,003
1881	2,297,131	19,703,074	3,326,916	3,433,593	3,820,946	3,989,834
1882	2,075,440	18,286,495	2,753,214	1,994,364	3,392,361	4,115,162
1883	1,296,713	18,851,231	2,821,537	3,451,506	4,292,217	3,937,259
1884	991,654	18,109,448	2,675,789	3,040,352	3,507,924	3,822,775
1885	1,097,198	18,133,677	2,883,937	3,175,738	3,551,783	4,156,095
1886	1,007,136	18,378,613	2,772,386	2,667,724	3,974,475	4,428,732
1887	1,019,032	15,149,419	2,071,457	3,569,402	2,719,151	3,599,185
1888	1,367,939	17,630,638	2,701,859	2,523,207	3,326,651	5,180,946
1889	1,436,163	17,632,645	3,242,522	2,847,622	3,909,822	4,854,001
1890	1,413,964	17,796,312	3,057,257	1,810,429	4,254,710	4,593,909
1891	1,401,127	18,833,576	3,310,459	3,036,586	4,349,464	4,342,852

IRISH AGRICULTURAL STATISTICS, 1876 - 1891

B. STATUTE ACREAGE UNDER CROPS, 1885 - 1891

SOURCE: HC 1892 [c. 6777], CXXXVIII.

	Wheat	Oats	Barley	Potatoes	Turnips	Hay
1885	71,017	1,328,869	179,133	797,292	296,984	2,034,768
1886	69,546	1,321,983	181,598	799,847	299,323	2,094,209
1887	67,181	1,315,055	162,169	796,939	300,123	2,143,818
1888	99,013	1,280,858	170,929	804,566	294,237	2,221,980
1889	89,745	1,238,952	185,783	787,234	297,913	2,187,522
1890	92,341	1,221,013	182,058	780,801	295,386	2,093,634
1891	80,870	1,215,396	177,966	753,332	300,326	2,059,529

C. THE NUMBER OF LIVESTOCK IN IRELAND, 1885 - 1991

SOURCE: HC 1892 [c. 6777], CXXXVIII.

	Horses & Mules	Asses	Cattle	Sheep	Pigs	Goats	Poultry
1885	576,430	197,170	4,228,851	3,478,056	1,269,092	264,437	13,850,532
1886	578,299	196,245	4,183,924	3,366,043	1,263,142	266,176	13,909,822
1887	587,234	199,512	4,157,404	3,377,826	1,408,456	271,729	14,460,643
1888	595,368	203,152	4,009,195	3,626,669	1,397,825	295,678	14,486,400
1889	604,102	206,236	4,094,174	3,789,187	1,380,670	303,933	14,856,517
1890	614,884	213,018	4,240,316	4,323,395	1,570,366	327,144	15,408,428
1891	621,479	216,268	4,448,511	4,722,613	1,367,712	336,337	15,276,128

IRISH AGRICULTURAL STATISTICS, 1876 - 1891

D. AGRICULTURAL PRICES, 1887 - 1891

SOURCE: HC 1890 - 1891, XCI; HC 1892 [c. 6777], CXXXVIII.

	Average Price for				
	1887	**1888**	**1889**	**1890**	**1891**
Wheat (pence per cwt)	31.3	34.8	32.3	33.5	35.8
Oats (pence per cwt)	24.4	26.7	28.4	30.4	34.2
Barley (pence per cwt)	30.0	32.9	32.9	32.7	35.3
Potatoes (pence per cwt)	11.3	12.7	13.4	15.2	17.8
Hay (pence per cwt)	12.6	10.0	8.8	9.8	16.3
Butter (pence per cwt)	452.9	449.2	480.0	432.8	517.7
Beef (pence per cwt)	235.4	267.1	281.8	279.4	–
Mutton (pence per cwt)	281.3	317.2	354.8	348.8	–
Fresh Pork (pence per cwt)	202.3	212.3	219.9	200.4	197.3
Wool (pence per lb)	4.2	4.2	3.9	4.0	4.0

Chapter Notes

Introduction

1. W. E. Vaughan, *Landlords and tenants in Ireland,* 1848-1904 (NP, 1984), p.32.
2. See, for instance, John E. Pomfret, *The struggle for land in Ireland, 1800-1923* (Princeton, 1930).
3. James S. Donnelly Jr, *The land and people of nineteenth century Cork: the rural economy and the land question* (London and Boston, 1975).
4. Paul Bew, *Land and the national question in Ireland, 1858-1882* (Dublin, 1978).
5. Samuel Clark, *Social origins of the Irish land war* (Princeton, 1979); 'The importance of agrarian classes: agrarian class structure and collective action in nineteenth century Ireland', in P. J. Drudy (ed), *Ireland, land, politics and people* (Irish Studies 2: Cambridge, 1982), pp. 11-36.
6. David Fitzpatrick, 'Class, family and rural unrest in nineteenth century Ireland', in Drudy (ed), *Ireland,* pp. 37-75.
7. Samuel Clark and James S. Donnelly Jr, *Irish peasants: violence and political unrest, 1780-1914* (Manchester University Press, 1983), pp. 421-2.
8. Vaughan, *Landlords and tenants,* p.32.
9. Paul Bew and Frank Wright, 'The agrarian opposition in Ulster politics, 1848-87, in Clark and Donnelly (ed), *Irish peasants,* pp. 192-229 and Brian M. Walker, 'The land question and elections in Ulster, 1868-86', in Ibid. pp. 230-268.
10. F. S. L. Lyons, 'John Dillon and the Plan of Campaign, 1886-1890', in *IHS,* XIV, No. 56 (September, 1965), pp. 313-47; *John Dillon: a biography* (London, 1968).
11. L. P. Curtis, Jr, *Coercion and conciliation in Ireland, 1880-92: a study in conservative unionism* (Princeton, 1963).
12. Emmet Larkin, *The Roman Catholic Church and the Plan of Campaign, 1886-1888* (Cork, 1978).
13. Anne B. Finnegan, 'The land war in south-east Galway, 1879-90', MA, UCG, 1974; Denis G. Marnane, 'Tipperary Town and the Smith-Barry estate, 1885-1895', MA, UCD, 1973; Laurence M. Geary, 'The land war on the Kingston estate, 1879-1888', MA, UCC, 1979.
14. Donnelly, *Land and people.*

Chapter 1: Land and Politics

1. PRO, Cabinet Minutes, CAB (1885) 37/16, 'The government of Ireland', 31 October, 7 December 1885.
2. Hansard, HC debs, 3S, CCCII, Cols 155-160, 21 January 1886.
3. B. L. Solow, *The land question and the Irish economy, 1870-1903* (Cambridge, Mass., and London, 1971), pp. 170-3.
4. *Report of the royal commission on the Land Law (Ireland) Act of 1881 and the Purchase of Land (Ireland) Act, 1885* [C. 4969], HC 1887, XXVI, 1, Paras 16, 18, 19. Hereafter cited as *Cowper Comm.*
5. *Special Commission Act, 1888; reprint of the shorthand notes of the speeches, proceedings and evidence taken before the commissioners appointed under the above-named act,* VII, 640. Hereafter cited as *Spec. Comm. Proc.*
6. *Cowper Comm.*, 8388.
7. *FJ*, 20 April 1886, Blunt to the editor, 19 April 1886.
8. Rents fixed under the Land Act of 1881.

9. *Cowper Comm.*, Para. 17.
10. SPO, CSORP 1888/26523.
11. *The Times*, 20 March 1886, Caird to the editor, 19 March 1886. Caird classified small farmers as those whose rent ranged from £1 to £20 a year and averaged £6. The average annual rent of the larger tenants was £56. The first class consisted of 538,000 holdings and the second of 121,000.
12. *Spec. Comm. Proc.*, VII, 640.
13. *Cowper Comm.*, 8388.
14. SPO, CSORP 1888/26523.
15. A. B. Cooke and J. R. Vincent (ed), *Lord Carlingford's journal: reflections of a cabinet minister, 1885* (Oxford, 1971), p. 136. Lord Carlingford had twice served as a minister under Gladstone, 1868–74 and 1880–5, and had played a leading part in the passage of the Land Acts of 1870 and 1881.
16. The *Nation*, 3 October 1885.
17. *FJ*, 12 October 1885.
18. PRO, Cabinet Minutes, CAB (1885) 37/16, 'Memorandum on the present state of Ireland', 26 September 1885.
19. *FJ*, 5 October 1885.
20. Ibid., 31 August 1885. The proposal to bank the rents was not new, as Healy indicated in his speech. In May 1881, the Galtee tenants of Mr Rice of Cork and the tenants of Captain Bowen of Bowen's Court, near Kildorrery, County Cork, banked their rents after their respective demands for an abatement had been refused. The Bowen tenants determined to 'fight it out to the bitter end'. *CE*, 2 and 12 May 1881.

    Both of these properties were contiguous to the estate of the Countess of Kingston at Mitchelstown, where a bitter rent strike was in progress. The entire locality was particularly tense at the end of April 1881 and remained so throughout the month of May. Lady Kingston had been granted nine ejectment decrees at the Midleton Quarter Sessions on 25 April and the first evictions took place at the end of May. The rent strike on the Kingston estate spread to virtually all neighbouring properties, irrespective of whether the rents were high or low. See Geary, 'Kingston estate'. Chapter 2.
21. *FJ*, 5 October 1885.
22. Ibid., 21 October 1885.
23. *UI*, 24 October 1885.
24. *Cowper Comm.*, 18,393. For similar resolutions, see *UI*, 7 November 1885, Caherline N.L., Co. Limerick; *UI*, 14 November 1885, Kilworth and Araglen N.L., Co. Cork; *UI*, 5 December 1885, Robeen N.L., Co. Mayo.
25. See, for instance, *Spec. Comm. Proc.* Report, 99–100; I, 288; IV, 271–2, 282.
26. See, for instance, the resolutions adopted by the County Grand Juries at the Spring Assizes in 1886, in *Irish public opinion in support of the legislative union* (ILPU, Dublin, 1886).
27. *The Times*, 6 January 1886, Trench to editor, 29 December 1885.
28. Ibid., 20 January 1886.
29. Ibid.
30. F. S. L. Lyons, *Charles Stewart Parnell* (London, 1977), pp. 292–4.
31. Hansard, HC debs, 3S, CCCII, Cols 155–7, 21 January 1886.
32. TCD Ms. 6500, John Dillon Papers, 'Irish parliamentary party minute book, 1886–1890', Vol. 1, 22 January 1886, 16 April 1886.
33. *FJ*, 22 February 1886. Archbishop William J. Walsh to Gladstone, 17 February 1886.

34. Ibid.
35. Lyons, *Parnell,* pp. 236-7.
36. Conor Cruise O'Brien, *Parnell and his party, 1880-1890* (Oxford, 1957), p. 165.
37. NLI Mss, Timothy C. Harrington Papers, 'National League letterbook, 1883-1888'. Harrington to J. J. MacMahon, Firies, Farranfore, 5 February 1886. See also Harrington to P. Griffin, Glin, 20 March 1886.
38. Ibid., Harrington to P. McCoy, Curramore, 19 June 1886.
39. Ibid., Harrington to Fr Brosnan, Abbeydorney, Tralee, 10 July 1886. Many letters similar to this one and the one to McCoy were written by Harrington to various branch secretaries during the spring and early summer of 1886.
40. *FJ,* 21 October 1885.
41. NLI Mss, Harrington Papers, Harrington to Rev. P. O'Connor PP, Farranfore, 19 November 1885.
42. HC 1893-94 [C.6935] XXXI, *Report of the commissioners appointed to inquire into the estates of evicted tenants in Ireland.* Hereafter cited as *ETC.*
43. See, for instance, *UI,* 24 October, 7, 14 November, 5 December 1885. On 19 December, an editorial entitled 'The rent struggle' stated that tenants were being forced 'in desperation' to lodge their rents with trustees, as the paper had previously suggested.
44. *Cowper Comm.,* 23,504.
45. PRO 30/6/67/26, Carnarvon Ms. quoted in A. B. Cooke and J. R. Vincent (ed), 'Ireland and party politics 1885-1887: an unpublished conservative memoir', in *IHS,* Vol. XVI, No. 164, September 1969, p. 470.
46. *Cowper Comm.,* 23,504.
47. Ibid., 16,404-408.
48. Ibid., 16,524-539.
49. Geary, 'Kingston estate', Chap. 3.

Chapter 2: Crisis

1. ILPU, *Irish public opinion in support of the legislative union* (Dublin, 1886).
2. *The Times,* 17 August 1886, A southern JP to the editor, 14 August 1886.
3. The special correspondent of *The Times, Letters from Ireland 1886,* (London, 1887).
4. *ETC,* 70-1.
5. *Cowper Comm.* 17,352-365.
6. Captain E. C. Hamilton, *The Woodford evictions (Dublin, 1886).*
7. *UI,* 24 July 1886, 'What next?', 7 August 1886, 'Is it rent or loot?', 14 August 1886, 'For bare life'.
8. *FJ,* 5 August 1886.
9. Lyons, *Parnell,* pp.358-9.
10. *FJ,* 10, 16 August 1886.
11. Ibid., 9 September 1886.
12. *The Times,* 14 August 1886.
13. Peter Marsh, *The discipline of popular government: Lord Salisbury's domestic statescraft 1881-1902* (Sussex and New Jersey, 1978), p.92.
14. *FJ,* 12 August 1886.
15. Ibid., 30 December 1886.
16. Hansard, HC debs, 3S, CCCVIII, Cols 65-70.
17. Ibid., Cols 564-5 and 587.

18. PRO, Buller Mss, WO 132/4A, 'Memorandum drawn up by the Secretary of State for War, W. H. Smith'.
    Despite his own and his hagiographer's protestations to the contrary, Buller's mission does not appear to have been particularly successful. C. H. Melville, *Life of General The Right Hon. Sir Redvers Buller* (2 Vols, London, 1923), 1, 318-9.
    At the Munster Winter Assizes, Chief Justice Morris said: 'The class of the cases from Kerry especially was of a startling character. Since the month of July there were three cases of murder, seventeen of threatening to murder, eight of attempted murder by shooting and there were cases of assault and arson and ... the maiming and injury of dumb beasts for the purpose of injuring indirectly their owners'. *The Times,* 3 December 1886.
    At the Spring Assizes in 1887, Justice O'Brien said that 'law, to a great extent, has ceased to exist' in Clare. *FJ,* 1 March 1887.
    In Kerry, he said, there existed 'a most extraordinary state of things, a state of things of an unprecedented description, nothing short, in fact, of a state of open war with all forms of authority and even, I may say without exaggeration, with the necessary institutions of a civilized life... a state of terror and lawlessness prevails everywhere'. *FJ,* 11 March 1887.
19. John Morley, *The life of William Ewart Gladstone* (3 Vols, London, 1911), 111, 274. See also Hansard, HC debs, 3S, CCCVIII, Col. 293.
20. Hansard, HC debs, 3S, CCCVIII, Cols 593-7.
21. *FJ,* 27 August 1886.
22. For details of the evictions, see Finnegan, 'The land war in south-east Galway', pp.81-5.
23. Hamilton, *Woodford evictions.*
24. Hansard, HC debs, 3S, CCCVII, Cols 1, 222-4.
25. Ibid., CCIX, Cols 984-1,000.
    Parnell wished to introduce a more extreme bill but was persuaded against it by Harcourt and Morley.
    Morley to Gladstone, 4, 10 September 1886, Gladstone Papers, BL, Add. Ms. 44,255, quoted in Thomas William Heyck, *The dimensions of British radicalism: the case of Ireland, 1874-95* (Urbana, Chicago, London, 1974), pp.187-8, n.26.
26. John Dillon, 'The coming winter in Ireland', in the *Nineteenth Century,* Vol.XX, No.CXVII, November 1886, pp.612-3.
27. Ibid.
28. Hansard, HC debs, 3S, CCCIX, Col.1207.
29. Lady Victoria Hicks Beach, *The Life of Sir Michael Hicks Beach, First Earl St. Aldwyn.* (2 Vols, London, 1932), 1, 283. Hereafter cited as Hicks Beach, *St. Aldwyn.*
30. Hansard, HC debs, 3S, CCCIX, Cols 1246-7; Dillion, 'Coming winter', pp.613-4.
31. *FJ,* 29 September 1886.
32. Ibid.
33. O'Brien, *Parnell and his party,* p.201.
34. *FJ,* 29 September 1886.

Chapter 3: Ultima Ratio Regum

1. SPO, *Special Commission 1888, Police Reports, Part 2, No. 7.*
2. *FJ,* 18 October 1886; F.S. L. Lyons, 'John Dillon and the Plan of Campaign, 1886-1890', in *IHS,* Vol.XIV, No 56, September 1965, pp. 315-6.

3. *UI,* 23 October 1886.
4. Ibid. It was reprinted as a special supplement on 20 November 1886. For the full text, see Appendix 1.
5. At a meeting of the central branch of the National League in Dublin on 30 December 1890, Harrington said: 'I had a part in the starting of the Plan of Campaign. I now tell you a secret. It was I who wrote the document of the Plan of Campaign'. *FJ,* 31 December 1890. See also William O'Brien, *Evening memories* (Dublin and London, 1920), p.158. Harrington's detractors snidely remarked that his admission came after the possibility of prosecution had long since passed. See, for example, *FJ,* 13 September 1892.
6. William O'Brien claimed that six or eight members were present, *Evening memories,* p.161. Dillon, Harrington and O'Brien himself are the only ones that can positively be identified.
7. R. Barry O'Brien, *The life of Charles Stewart Parnell, 1846-1891* (2 Vols, London, 1899), 11, 170-171; Lyons, *Parnell,* p.362.
8. T. M. Healy, *Letters and leaders of my day* (2 Vols, London, 1928), 1, 226; A. J. Kettle, *The material for victory* (Dublin, 1958), p.73.
9. J. F. Lalor, 'The faith of a felon' in *The Irish Felon,* 8 July 1848.
10. *FJ,* 31 August 1885, 5 October 1885. See above pp.8-9
11. *The Nation,* 10 October 1885.
12. *DE,* 30 October 1886.
13. SPO, CSORP 1888/26521, Clonmel, Co. Tipperary, 6 November 1887.
14. SPO, CSORP 1891/28047.
15. SPO, CSORP 1888/26521, Rathgormac, Co. Waterford, 1 January 1888. See also *FJ,* 2 January 1888, Carrick-on-Suir, 1 January 1888; SPO, CSORP 1887/10081, Carrickmacross, Co. Monaghan, 29 May 1887 and TCD Mss, Davitt Papers, DP/AA6, pp.15-16.
16. *FJ,* 8 November 1886, Kenagh, Co. Longford, 7 November 1886; *FJ,* 4 January 1887, Gorey, 3 January 1887.
17. *UI,* 11 December 1886.
18. *FJ,* 22 November 1886, Keadue, Co. Roscommon, 21 November 1886. See *FJ,* 21 March 1887, Crilly at Clonkeen, Co. Roscommon, 20 March 1887; SPO CSORP 1887/17196, Crilly at Liseragash, Co. Roscommon, 15 April 1888; Hansard, 3S, Vol.312, 1640, Sheehy at Castlerea, 5 December 1886; SPO CSORP 1888/26521, J. Redmond at Claremorris, 19 January 1887; *UI,* 30 October 1886, 6 November 1886, W. Redmond at Castleblakeney, 24 October 1886 and at Loughrea 31 October 1886.
19. Hansard, 3S, Vol.312, 1639.
20. SPO, CBS, DICS, South-Eastern Division, 1887-1895, No.3, 'List of Nationalist Members of Parliament with detailed biographical notes'.
21. ILPU, *The Plan of Campaign, illustrated: an account of the Massereene estate.* (Dublin, 1888), p.15.
22. SPO, CBS, 1890 1209/S.
23. SPO, Police and Crime Records, 1887-1917, Crime Branch Special. Divisional Commissioners and County Inspectors Monthly Confidential Reports, 1887-1898, Nos 1-3, Western Division, November 1898. Hereafter cited as SPO, P and CR, CBS, Nos 1-3.
24. SPO, CSORP 1888/17196, Liseragool, 16 April 1888.
25. HC 1888 (C. 5401) LXXXIII. 11. *Wilfrid Scawen Blunt v John Byrne in the High Court of Justice in Ireland, the judge's charge.* Hereafter cited as *Blunt v Byrne.*
26. SPO, CBS, 1892 5528/S.

27. Ibid.
28. 'The papal rescript', in *The Westminster Review,* Vol. CXXIX, June 1888, pp.693-70.
29. William J. Duffy Papers, Fahy to Duffy, 30 October 1886.
30. SPO, CSORP 1888/26521, Fr Donnellan CC at Finnegan's Cross, Co. Armagh, 18 September 1887.
31. Ibid., Fr Lane CC at Ballinadee, Co. Cork, 30 October 1887.
32. Ibid., Fr Woods CC at Kells, Co. Meath, 4 December 1887.
33. SPO, CSORP 1887/10081, Fr L. Gaughran at Carrickmacross, Co. Monaghan, 29 May 1887.
34. SPO, CSORP 1888/26521, Fr Rock CC at Collon, Co. Louth, 29 May 1887.
35. SPO, CSORP 1891/28047.
36. SPO, CSORP 1888/26521, Fr Kennedy at Meelin, Co. Cork, 14 August 1887.
37. Ibid., Fr Ryan at Herbertstown, 1 November 1887.
38. SPO, CBS, DICS, No. 8, 19 March 1889.
39. PRO, CO 903/1, Irish Crime Records, 1885-1892, Intelligence Notes XII, 'Memorandum on Olphert estate', 17 June 1889.
40. SPO, CSORP 1889/11318, McFadden at Derryart, 4 January 1888.
41. SPO, CSORP 1891/28048, Belfast, 10 April 1888.
42. Ibid., Ardnagappery, 21 September 1887.
43. *FJ,* 25 November 1887; *UI,* 3 December 1887.
44. SPO, CSORP 1891/28048, Derrybeg chapel yard, 4 December 1887.
45. PRO, CO 903/1, Irish Crime Records, 1885-1892, Intelligence Notes XII, 'Memorandum on Olphert estate', 17 June 1889.
46. SPO, CSORP 1891/28048, Derrybeg chapel yard, 1 January 1888; SPO, CSORP 1889/11318, Doe, 4 January 1888.
47. SPO, CSORP 1891/28048.
48. O'Brien had just completed a three-months prison sentence, imposed for speeches delivered at Mitchelstown in August 1887.
49. SPO, CSORP 1891/28408.
50. SPO, Irish Crime Records, 1887-1892, Prosecutions under the Criminal Law and Procedure Act, 1887. Hereafter cited as CLPA Prosecutions.
51. Archbishop Logue to Tobias Kirby, Rector of the Irish College, Rome, 15 February 1888, quoted in Larkin, *Catholic Church and Plan of Campaign,* pp.186-7.
52. SPO, CSORP 1889/11318.
53. SPO, CSORP 1891/28048.
54. SPO, CSORP 1889/11318.
55. PRO, CO 903/, 'Olphert estate'.
56. Leon Ó Broin, *The prime informer: a suppressed scandal* (London, 1971), pp. 45-6; Desmond Murphy, *Derry, Donegal and modern Ulster, 1790-1921* (Derry, 1981), pp187-90; Sir John Ross, *The years of my pilgrimage* (London, 1924), pp.54-7.
57. SPO, CBS, DICS, No.8, 21 November 1889.
58. Murphy, *Derry,* pp.189-90.
59. Wilfrid Scawen Blunt, *The land war in Ireland, being a personal narrative of events* (London, 1912), pp.320-1; p.71.
60. *FJ,* 20 December 1887, O'Dwyer to the editor, 19 December 1887.
61. Ibid., 21 December 1887, Healy to the editor, 20 December 1887.
62. PRO, CO 903/1, Irish Crime Records, 1885-1892, Intelligence Notes XII, 'Memorandum on the Kenmare estate', 17 June 1889.

63. Patrick J. Corish (ed), 'The Kirby Papers, Irish College, Rome. Guide to material of public and political interest, 1884-1894, with addenda 1852-1878', in *Archivium Hibernicum*, XXXII, 1974, Most Rev. John Power to Kirby, 4 May 1888, No.205, p.15.
64. Ibid., John MacEvilly, Archbishop of Tuam, to Kirby, 29 July 1888, No.287, p.17. See also Walsh to Kirby, 9 March 1887, No.139, p.11.
65. Ibid., John McCarthy, Bishop of Cloyne, to Kirby, 9 January 1889, No.10, p.20 and 11 July 1888, No.264, p.17.
66. SPO, P and CR, CBS, South-Western Division, November 1887, May 1888.
67. Larkin, *Catholic Church and Plan of Campaign*, pp.4-10.
68. Thomas Brett, *Life of the Most Reverend Dr Patrick Duggan, Bishop of Clonfert* (NP, 1912), pp.173-4, Duggan to Dean Ryan, 15 December 1886.
69. *Leinster Leader*, 19 February 1887.
70. *Midland Tribune*, 24 February 1887.
71. Blunt, *Land war*, p.44.
72. *FJ*, 9 May 1889.
73. Ibid., 24 May 1889, Lord Massereene to the editor, 23 May 1889; T. W. Russell, *Disturbed Ireland: the Plan of Campaign estates* (London, 1889), the Olphert estate, p.52; the Brooke estate, p.142; the Lansdowne estate, p.121; NLI Ms. 4275, Letter Book of Arthur Moore, agent on the estate of H. J. Murray-Stewart, Ardara, Co. Donegal, Moore to Murray-Stewart, 13 January 1888.
74. SPO, CSORP 1886/22884.
75. SPO, CSORP 1887/1101.
76. *FJ*, 19 November 1886.
77. SPO, CSORP 1887/1519, 'Rents and the Plan of Campaign in the Western Division', 24 January 1887.
78. *FJ*, 8 November 1886, Drum, Co. Roscommon, 7 November 1886.
79. SPO, CSORP 1888/26521, Clonmel, Co. Tipperary, 6 November 1887.
80. *FJ*, 24 January 1887, Maryborough, Queen's County, 23 January 1887.
81. *UI*, 6 November 1886, Loughrea, Co. Galway, 31 October 1886.
82. *FJ*, 19 November 1886, Dillon at Portumna, Co. Galway, 18 November 1886.
83. SPO, CSORP 1888/262521, Collon, Co. Louth, 29 May 1887.
84. PRO CO 903/1, Irish Crime Records, 1885-1892, Intelligence Notes IV, Meeting at Herbertstown, end of April 1888.
85. SPO, CSORP 1888/17196.
86. Ibid.
87. SPO, P and CR, CBS, Nos 1-3.
88. NLI Ms. 22231, George C. Mahon Papers, John Ross Mahon to George C. Mahon, 25 April 1887.
89. *The Times*, 5 November 1885, Richard Bagwell to the editor, quoted in Cork Defence Union, *Boycotting in the County of Cork* (Cork, 1886). For an example of boycotting on a Plan of Campaign estate, see Geary, *Kingston estate*, pp.68-71.
90. *The Times*, 26 October 1886, Lord Clanricarde to the editor, 24 October 1886.
91. West Yorkshire PRO, Clanricarde Mss, Bundle 84, Joyce to Clanricarde, 18 November 1886.
92. *FJ*, 22 November 1886.
93. Ibid., 19 November 1886. For a flamboyant account, see O'Brien, *Evening memories*, pp.171-4.
94. *The Times*, 24 November 1886, Clanricarde to the editor, 22 November 1886.
95. West Yorkshire PRO, Bundle 84, Joyce to Clanricarde, 18 November 1886.
96. *FJ*, 3, 10 December 1886.

97. Ibid., 1, 2, 9, 10 December 1886.
98. SPO, CSORP 1887/1519.
99. William O'Brien Mss, UCC, Box AA, Crilly to O'Brien, 3 December 1886.
100. SPO, P and CR, CBS, DICS, South-Eastern Division, 1887-1895.
101. SPO, CSORP 1887/1022.
102. SPO, CSORP 1886/22884.
103. SPO, CSORP 1887/1101.
104. SPO, P and CR, CBS, Western Division, February 1888.
105. *FJ,* 19 November 1886. See also *FJ,* 8 November 1886, O'Brien at Drum, Co. Roscommon; *Leinster Leader,* 8 October 1887, O'Brien at Luggacurran, Queen's County; *FJ,* 22 November 1886, Deasy at Drumlish, Co. Longford.
106. *The Plan of Campaign, illustrated: an account of the Ponsonby, Kingston, Lansdowne, O'Grady and Brooke estates* (ILPU, Dublin, 1888), p.18.
107. *The Plan of Campaign, illustrated: an account of the Massereene estate* (ILPU, Dublin 1888), pp.7-8.
108. SPO, CSORP 1888/17196.
109. FJ, 4 April 1887, Tynagh, Co. Galway.
110. Ibid., 22 November 1886, Drumlish, Co. Longford.
111. Ibid., 12 April 1887, Draperstown, Co. Derry.
112. *Blunt v Byrne,* p.21-22, Kylebeg, 21 November 1886.
113. SPO, CLPA Prosecutions.
114. *Blunt v Byrne,* p.21-22, Kylebeg, 21 November 1886.
115. PRO, CO 903/1, Irish Crime Records, 1885-1892, Intelligence Notes XI, 'The Plan of Campaign. Its origin, progress and present aspect', 26 January 1888.
116. Hansard, 3S, Vol.322, Col.1,340.
117. *ETC,* 14940-54.
118. See Appendix 2 and Plan of Campaign distribution maps.
119. Irish Land Commission, Record A.83.
120. SPO, CSORP 1888/7644.
121. SPO, CSORP 1891/24048; BL, Balfour Papers, Add. Ms. 49688, Henry Robinson to Balfour, 1 July 1887; *Cowper Comm.,* 24233-6.
122. *Cowper Comm.,* Report 27-30.
123. SPO, CSORP 1891/24048.
124. BL, Balfour Papers, Add. Ms. 49688, Robinson to Balfour, 1 July 1887.
125. *Cowper Comm.,* 24233-6; *FJ,* 22 May 1888, John Murtagh to the editor, 16 May 1888.
126. SPO, CSORP 1891/24048; *ETC,* 635-778.
127. SPO, CSORP 1891/24048.
128. Alfred E. Pease MP, *Glensharrold in 1888* (London, 1888), Appendix 1, pp.42-5.
129. SPO, CSORP 1888/2222.
130. *Midland Tribune,* 20 December 1889, quoted in PRO, Co 903/1, Irish Crime Records, 1885-1892, Intelligence Notes IV.
131. ILPU, *The Plan of Campaign, illustrated: the Ponsonby estate,* pp.22-3.
132. PRO, Cabinet Minutes, CAB (1889) 37/24. 'Observations by the Deputy Inspector General of the RIC on certain matters connected with the condition of Down, Armagh and Louth in April 1889', H. Thynne, 6 April 1889.
133. *FJ,* 22 May 1888; *UI,* 26 May 1888.
134. *FJ,* 6 December 1886.
135. SPO, CSORP 1888/26521, Kells, 4 December 1887.
136. BL, Balfour Papers, Add. Ms. 49688, Robinson to Balfour, 1 July 1887.

137. Ibid., Add. Ms. 49807; *UI,* 26 May 1888; SPO, P and CR, 1887-1917, CBS, Nos 1-3, Western Division; SPO, CSORP 1888/7644, Brown to Balfour, 9 March 1888.
138. SPO, P and CR, CBS, DICS, Western Division, September 1888.
139. *Cowper Comm.,* 15485-15512.
140. L. P. Curtis Jr, 'Incumbered wealth: landed indebtedness in post-Famine Ireland' in *The American Historical Review,* Vol.85, No.2, April 1980, pp.340-2.
141. Mortgage ledgers of the Representative Body of the Church of Ireland, Church of Ireland House, Church Avenue, Rathmines, Dublin 6.
For Massereene, see PRO, Cabinet Minutes, CAB (1889) 37/24 and Sir Alfred E. Pease, *Elections and recollections* (London, 1932), p.199.
For Lady Kingston's financial troubles, see Geary, 'Kingston estate'.
142. Irish Land Commission Records, CDB/3159.
143. At the General Synod of the Church of Ireland on 6 May 1886, the Right Hon. R. R. Warren, on behalf of the Finance Committee, stated that no further church money would be invested on the security of Irish land, *Journal of the General Synod, 1886,* XCVIII. From May 1886 to the end of 1891, roughly the period of the Plan of Campaign, the Representative Church Body refused all applications for loans upon mortgage of land, *Report of the Representative Body of the Church of Ireland* (Dublin, 1892), p.12.
144. Maynooth College Records, in the possession of the Rev. President.
145. *IT,* 28 July 1893.
146. Pease, *Glensharrold in 1888,* pp.8-10; ILPU, *The Catholic Bishop of Limerick on the Glensharrold evictions* (Dublin, 1890); *FJ,* 29 May 1888.
147. *FJ,* 30 March 1889.
148. BL, Balfour Papers, Add. Ms. 49688, Robinson to Balfour, 1 July 1887.
149. Terence H. O'Brien, 'Lord Milner's Irish Journal, 1886' in *History Today,* Vol.XIV, January 1964, p.46.
150. Ibid.
151. PRO, CO 903/2, Irish Crime Records, 1887-1892, Miscellaneous Notes, Series XIII, June 1891.
152. Blunt, *Land war,* p.429.
153. NLI Ms. 22461, The O'Grady Papers.
154. Ibid., statement showing the loss caused to The O'Grady by the action of his tenants in joining the Plan of Campaign.
155. George Pellew, *In castle and cabin, or talks in Ireland in 1887* (New York and London, 1888), p.151.
156. *Kildare Observer,* 20 November 1886.
157. Lord Newton, *Lord Lansdowne: a biography* (London, 1929), pp.24-5; HC 1876 [C. 1492] LXXX. 35. *Return of the owners of land of one acre and upwards . . . in Ireland.* For details of mortgages and family charges, see Irish Land Commission Records, LCI/LC 797 F.S.I.
158. Newton, *Lansdowne,* pp.45-51, Lansdowne to his mother, 23 March 1887, 31 July 1887, 13 October 1887, beginning of December 1887, 8 February 1888.
159. ILPU, *The Plan of Campaign, illustrated,* p.14.
160. *FJ,* 30 March 1887.
161. Conversely, the two most obdurate opponents of the Plan were Lord Clanricarde and A. H. Smith-Barry, very wealthy men who could afford, and were prepared, to sustain heavy losses in their attempts to crush what they saw as an anarchic conspiracy.

162. PRONI, Saunderson Papers, T. 2996/3/19, Col Edward Saunderson to Lord Randolph Churchill, 12 January 1887.
163. Pease, *Elections and recollections,* p.133.
164. *Cowper Comm. Report,* 12.
165. *Cowper Comm.,* 7035.
166. *UI,* 12 March 1887.
167. SPO, CSORP 1887/218, 1887/1518.
168. David W. Miller, *Queen's rebels: Ulster loyalism in historical perspective* (Dublin and New York, 1978), p.115.
169. *Cowper Comm.,* 5069.
170. Ibid., 10376.
171. Ibid., 3994.
172. Ibid., 6855-9.
173. Clark and Donnelly (ed), *Irish peasants,* pp.149-150.
174. R. F. Foster, 'To the northern counties station: Lord Randolph Churchill and the prelude to the orange card' in F.S.L. Lyons and R.A.J. Hawkins (ed), *Ireland under the Union: varieties of tension. Essays in honour of T. W. Moody* (Oxford, 1980), pp.275-6.
175. Quoted in T. W. Russell, *Ireland and the Empire: a review, 1800-1900* (London, 1901), p.146.
176. *Cowper Comm.,* 9316.
177. Ibid., 6828.
178. *The Times,* 13 December 1886.
179. *IT,* 21 January 1891.
180. See above pp.25-6.
181. *DE,* 13 December 1886.
182. SPO, CSORP 1887/22054.
183. Hansard, 3S, Vol.313, Cols 1624-5.
184. *Dundalk Democrat,* 4 December 1886, cited in Peadar Livingstone, *The Monaghan story: a documented history of the County Monaghan from the earliest times to 1976* Enniskillen, 1980), pp.348-9.
185. Livingstone, *Monaghan,* p.349.

Chapter 4: 'Mickey-the-Botch'

1. Salisbury Mss, D/9/161, Hicks Beach to Salisbury, 25 July 1886; also in Hicks Beach, *St Aldwyn,* 1, 274-5.
2. Curtis, *Coercion and conciliation,* pp. 133-7; A. B. Cooke and A. P. W. Malcomson (ed), *The Ashbourne Papers, 1869-1913. A calendar of the papers of Edward Gibson, 1st Lord Ashbourne* (PRONI, 1974), p.27.
3. St Aldwyn Mss, PCC/45; PRO, Buller Mss, WO 132/4A; Melville, *Buller,* 1, 310-15, Buller to Hicks Beach, 30 August, 4, 5, 18 September, 15 November 1886.
4. St Aldwyn Mss, PCC/17; Hicks Beach, *St Aldwyn,* 1, 289-90, Hicks Beach to Buller, 18, 22, 26 September 1886.
5. *Cowper Comm.,* 15,826. See also appendix D paper one, Lord Lansdowne to J. Townsend Trench, 21 October 1886.
6. PRO, Buller Mss, WO 132/4A, Beach to Buller, 14 October 1886.
7. *ETC,* 63; also in *UI,* 17 December 1887, and Hicks Beach, *St Aldwyn,* 1, 292-4.
8. Salisbury Mss, D/9/162; also in St Aldwyn Mss, PCC/31, and in Hicks Beach, *St Aldwyn,* 1, 294-5.

9. *UI,* 23 October 1886; O'Brien, *Evening memories,* p. 166 n.1.
10. *FJ,* 25, 28 October 1886.
11. Ibid., 8 November 1886.
12. Ibid.
13. Ibid., 1 November 1886.
14. John J. Clancy (ed), *Mr Dillon and the Plan of Campaign: being Mr Dillon's speech before the Queen's Bench, with appendices* (Irish Press Agency, London, 1887), pp.13-18.
15. *FJ,* 19 November 1886.
16. Ibid., 10 November 1886. See also *The Times,* 9 November 1886, Buller to the editor, 6 November 1886 and *The Times,* 15 November 1886, Hicks Beach at Bristol, 13 November 1886.
17. *FJ,* 10 November 1886.
18. *Cowper Comm.,* 16,454–97. See also Melville, *Buller,* 1, 314-15, Buller to Hicks Beach, 15 November 1886. Within a fortnight it was announced that Buller was to succeed Sir Robert Hamilton as Under Secretary at Dublin Castle. In view of the appointment, Hicks Beach felt that it would have been better had Buller not given evidence before the Cowper Commission but, as he had already done so, the Chief Secretary said that he knew of no reason why the government should trust Buller less for having said publicly what he had frequently told Hicks Beach in private. PRO, Buller Mss, WO 132/4A; Melville, *Buller,* 1, 320, Hicks Beach to Buller, 26 November 1886.
19. *The Times,* 15 November 1886.
20. Ibid., 18, 23 November, 2 December 1886.
21. For details of the rent collections, see Chapter 3, pp. 37-9.
22. *Blunt v Byrne.*
23. Hicks Beach to Salisbury, 30 November 1886, quoted in Lyons, 'John Dillon and the Plan of Campaign, p.318.
24. SPO, CSORP 1886/22361. Head Constable Patrick Kielty reported that, at Loughrea on 9 December, tenants were given receipts which were 'partly printed, with the name of the tenant and the amount paid put in pencil but there was no name signed'.
25. PRO, Cabinet Minutes, CAB (1887) 37/19, 'The Plan of Campaign', a memorandum drawn up by Sir Michael Hicks Beach, 13 January 1887.
26. *UI,* 4 December 1886.
27. Larkin, *Catholic Church and the Plan of Campaign,* pp.4-10.
28. *UI,* 4 December 1886; O'Brien, *Evening memories,* p.159.
29. See Appendix 2, 'Plan of Campaign Estates'.
30. *FJ,* 9 December 1886.
31. *UI,* 11 December 1886.
32. *FJ,* 3 December 1886.
33. NLI Ms.8574 (22), T. Harrington Papers, Rev. D. Harrington, St Michael's College, Listowel, to Harrington, 5 December 1886.
34. *FJ,* 6 December 1886.
35. See, for instance, his speech at Castlerea on 5 December, above, p.25.
36. Clancy, *Dillon,* pp.18-19.
37. *FJ,* 15 December 1886.
38. Ibid., 17 December 1886.
39. Ibid., 20 December 1886.
40. *The Times,* 20 December 1886.
41. *FJ,* 22 December 1886.

42. Ibid., 23, 24 December 1886.
43. *IT,* 31 December 1886 — 12 January 1887.
44. O'Brien, *Evening memories,* p.203.
45. *The Times,* 11 January 1887, *CC,* 1 February 1887.
46. Reginald Lucas, *Colonel Saunderson MP* (London, 1908), p.124.
47. *DE,* 16 February 1887.
48. PRO, Cabinet Minutes, CAB (1887) 37/19.
49. Curtis, *Coercion and conciliation,* p.166.
50 Hansard, HC debs, 3S, Vol.310, Col.4.
51. See, for example, *FJ,* 14, 17, 20, 23, 24, 25, 27, 31 January, 7, 14 February 1887.
52. PRO, Cabinet Minutes, CAB (1887) 37/19.
53. See Appendix 2, 'Plan of Campaign Estates'.
54. *FJ,* 10 January 1887.
55. O'Brien *Evening memories,* p.208; *FJ,* 14 February 1887.
56. *FJ,* 10 January 1887.
57. Hicks Beach, *St Aldwyn,* 1, 299, Hicks Beach to Salisbury, 15 December 1886.
58. See Appendix 1, 'A Plan of Campaign'.
59. NLI Ms.22461, The O'Grady Papers, circular from The O'Grady to his tenants, 13 August 1887.
60. *FJ,* 14-25 February 1887.
61. Hicks Beach, *St Aldwyn,* 1, 312-3.
62. Pease, *Elections and recollections,* p.161.
63. Hansard, HC debs, 3S, Vol.311, Col.1107.
64. Ibid., Col.1182.
65. Pease, *Elections and recollections,* pp.161-2.
66. Hicks Beach, *St Aldwyn,* 1, 312-6.
67. A Youghal fisherman named Hanlon received fatal injuries from a police bayonet during a public demonstration on the day after Hicks Beach's 'worse than batons' speech.
68. *FJ,* 7 March 1887.
69. Ibid.
70. Hicks Beach, *St Aldwyn,* 1, 316-20.
71. *FJ,* 7 March 1887.
72. Salisbury Mss, D/9/186, Salisbury to Hicks Beach, 28 February 1887.
73. T. Maguire, Professor of Moral Philosophy, Trinity College, Dublin, in *The Union,* 5 March 1887.
74. See, for example, Sir Henry Robinson, *Memories: wise and unwise* (London, 1923), p.90.
75. T. M. Healy, *Letters and leaders,* 1, 269.

Chapter 5: 'Tiger Lily'

1. Salisbury Mss, correspondence with Sir Michael Hicks Beach, 1887-1891, folios 253-322, Hicks Beach to Salisbury, 5 March 1887.
2. SPO, P and CR, CBS, 1887-1917, Nos 1-3, Western Division.
3. Ibid., No. 4, South-Western Division.
4. H.C. 1887 [C. 4955] LXVIII.38. *Return of the number of agrarian outrages committed in Ireland which were reported to the Inspector General of the Royal Irish Constabulary during the quarter ended 31 December 1886.* H. C. 1887 [C. 5030] LXVIII.37, *quarter ending 31 March 1887.*
5. Pease, *Elections and recollections,* pp.165-6. See also *The Times,* 21 March 1887.
6. Healy, *Letters and leaders,* 1, 270.

7. *FJ*, 12 March-28 May 1887.
8. *UI*, 25 June 1887.
9. SPO, CSORP 1887/1906.
10. Curtis, *Coercion and conciliation*, p.155. See also Henry Norman, *Bodyke: a chapter in the history of Irish landlordism* (London, 1887), pp.11-12.
11. SPO, P and CR, 1887-1917, CBS, No. 4, May 1887.
12. For details of the evictions, see *FJ*, 2-18 June 1887.
13. Ibid., 3 June 1887.
14. Ibid., 8 June 1887.
15. Ibid., 9 June 1887.
16. Ibid.
17. Quoted in *UI*, 11 June 1887.
18. Hansard, HC debs, 3S, Vol.316, Cols 278-367.
19. BL, Balfour Papers, Add. Ms.49826, Balfour to Buller, 11 June 1887.
20. FJ, 5-19 July 1887.
21. Ibid., 16 July 1887.
22. Ibid., 18 July 1887.
23. Ibid., Walsh to editor, 16 July 1887.
24. BL, Balfour Papers, Add. Ms.49807, Turner to Buller, 29 July 1887.
25. Curtis, *Coercion and conciliation*, pp.180-1.
26. Ibid., pp.337-43.
27. Salisbury Mss, D/9/189, Salisbury to Hicks Beach, 3 July 1887.
28. Cooke and Malcomson (ed), *Calendar of the Ashbourne Papers*, p.188.
29. *The Times*, 3 Feb. 1891, O'Brien to the editor, 31 January 1891. See also *FJ*, 3 October 1892.
30. H.C. 1893-94 (215) LXXIV. Part II. *Return of inquiries held under section one of the Criminal Law and Procedure (Ireland) Act 1887, showing (i) county and locality in which each inquiry was held, (ii) nature of the offence which was the subject of inquiry and (iii) result of each such inquiry*.
31. SPO, CSORP 1888/17196.
32. Ibid. and P and CR, 1887-1917, CBS, Nos. 1-3, November 1888.
33. SPO, CSORP 1888/17196.
34. I have discussed the Mitchelstown Massacre at some length elsewhere. See Geary, *Kingston estate*, Chapter 4.
35. Ibid.
36. BL, Balfour Papers, Add. Ms.49688, Salisbury to Balfour, 26 October 1887.
37. H.C. 1889 (158) LXI 515 *Criminal Law and Procedure (Ireland) Act, 1887. Proceedings against members*, and SPO, CLPA Prosecutions.
38. SPO, CLPA Prosecutions.
39. *The Times*, 10 March 1887, McCarthy to Keller, 7 March 1887. *The Times* commented: 'Such a communication as this from a high ecclesiastical dignitary to a clergyman who is subject to his discipline and bound to follow his guidance is one of the strongest proofs which could be given of the spread of anarchic ideas and the decline of moral power to control them'.
40. SPO, CLPA Prosecutions.
41. *The Times*, 10 November 1887. See also *The Times*, 12 January 1888, Salisbury at Liverpool.
42. Ridgeway MSS, Buller to Ridgeway, 15 November 1887.
43. *The Times*, 1 November 1887.
44. SPO, P and CR, 1887-1917, CBS, Nos 1-3 and 4, November 1887.

45. BL, Balfour Papers, Add. Ms.49689, Ferguson to Peter O'Brien, 25 February 1888. See also Add. Ms.49688, 'Memorandum on the present state of the South and West of Ireland', compiled by the Hon. Keith Falconer, Special Commissioner of Income Tax, 16 December 1887.
46. Ibid., Add. Ms.19822.
47. SPO, CLPA Prosecutions.
48. PRO, CO 903/2, Irish Crime Records, 1887-92, Misc. Notes, Series III.

Chapter 6: The Pope One Day, Parnell The Next

1. *FJ,* 9 April 1888.
2. Blunt, *Land war in Ireland,* pp.418-19.
3. The full text of the papal rescript is in Larkin, *Catholic Church and Plan of Campaign,* pp.201-2. Larkin extensively analyses the intrigue that led to the Pope's condemnation of the Plan and boycotting.
4. *FJ,* 28 April 1888.
5. Blunt, *Land war in Ireland,* p.420; O'Brien, *Evening memories,* pp.348-50.
6. O'Brien, *Parnell and his party,* p.217; Lyons, *Dillon,* pp.93-4.
7. *FJ,* 5 May 1888.
8. Ibid., 8 May 1888.
9. George Bernard Shaw, *The matter with Ireland,* edited with an introduction by David H. Greene and Dan H. Lawrence (London, 1962), pp.18-20.
10. For the reaction of the hierarchy to the rescript, see Larkin, *Catholic Church and Plan of Campaign,* Part 11.
11. *FJ,* 9 May 1888.
12. Ibid., 18 May 1888.
13. Ibid., 21 May 1888.
14. Ibid., 26 May 1888.
15. Ibid., 28 May 1888.
16. Ibid.
17. Ibid., 30 May 1888.
18. Ibid., 31 May 1888.
19. Ibid., 12 June 1888.
20. See Appendix 2, 'Plan of Campaign Estates'.
21. SPO, P and CR, CBS, DICS, South-Eastern Division, September 1888.
22. Ibid., CBS No. 4, South-Western Division, June 1888, January 1889.
23. BL, Balfour Papers, Add. Ms.49821, 'Memorandum on the Pope's action in Ireland', 19 March 1889.
24. *IT,* 6, 10 May 1890, Humphries to the editor, 9 May 1890.
25. Corish (ed), *Archivium Hibernicum,* XXX II, 1974, Kirby Papers. See also Archbishop Walsh to Kirby 3 July 1888, 4 February 1889.
26. *FJ,* 9 May 1888.
27. See above p.25.
28. *The Times,* 8 December 1886.
29. Ibid., 9 December 1886.
30. *DE,* 10 December 1886, *The Tablet,* 11 December 1886.
31. See, for example, Gladstone at the Memorial Hall, London, 29 July 1887, quoted in Lord Everley, *Gladstone and Ireland* (London, 1912) pp.327-8. See also Hansard, HC debs, 3S, Vol.310, Cols 108-10, 27 January 1887 and Vol.322, Cols 770-78, 17 February 1888.
32. Gladstone to Morley, 8 December 1886, quoted in Morley *Gladstone,* iii, 280.
33. Morley *Gladstone,* iii, 280.

34. O'Brien, *Evening memories,* pp.176-90; Morley *Gladstone,* iii, 280.
35. *FJ,* 22 December 1886.
36. *Pall Mall Gazette,* 22 December 1886.
37. Blunt, *Land war in Ireland,* p.198.
38. Lyons, *Parnell,* pp.218-21.
40. BL, Balfour Papers, Add. Ms.49808, Ridgeway to Balfour, 10 May 1888.
41. TCD Ms.6736, John Dillon Papers, correspondence with William O'Brien, 1887-1892.
42. Lyons, *Dillon,* pp.92-3 and 101.
43. See *FJ,* February-September 1888.
44. William O'Brien Papers, UCC, Box AB.
45. SPO, CSORP 1888/2774.
46. BL, Balfour Papers, Add. MS.49802, Balfour to the Marquess of Londonderry, 23 May 1888.
47. SPO, P and CR, 1887-1917, CBS, South-Western Division, May 1888.
48. BL, Balfour Papers, Add. Ms.49808, Ridgeway to Balfour, 10 July 1888.
49. *FJ,* 19 July-1 August 1888.
50. Russell, *How the Irish members got into gaol,* pp.58-60; *FJ,* 17-21 August 1888; SPO, CLPA Prosecutions; *ETC,* p.313, James E. Byrne to Lister Drummond, 1 December 1892.
51. BL, Balfour Papers, Add. Ms.49809. Also in PRO, Cabinet Minutes, CAB (1888) 37/32, No. 33.
52. BL, Balfour Papers, Add. Ms.49826, Balfour to Ridgeway, 12 May 1888.
53. *FJ,* 10-14 May, 21 June 1888. See also H.C. 1889 (158) LXI 515.
54. SPO, CLPA Prosecutions. See also O'Brien, *Evening memories,* pp.347-8.
55. TCD Ms.6736, John Dillon Papers, W. O'Brien to Dillon, 1 July 1888.
56. M. Mc. D. Bodkin, *Recollections of an Irish judge: press, bar and parliament* (London, 1914), pp.160-2. See also H.C. 1889 (158) LXI 515.
57. For the Irish leader's relationship with Rhodes, see Lyons, *Parnell,* pp.442-4, 587-9.
58. TCD Ms.6736, John Dillon Papers, O'Brien to Dillon, 14, 16 July 1888.
59. BL, Balfour Papers, Add. Ms.49827, Balfour to Ridgeway, 28 June, 30 August 1888.
60. Ibid., Add. Ms.49689, Salisbury to Balfour, 20 September 1888.
61. TCD Ms.6559, John Dillon Papers, journal entry, 28 October 1888.
62. Ibid., 5 November 1888.
63. Ibid., 6745, Dillon to Parnell, 14 January 1889.
64. Ibid., Parnell to Dillon, 26 January 1889.
65. See, for example, O'Brien to Dillon, 1, 16 July 1888, TCD Ms.6736.
66. Ibid., no date, probably mid-late July 1888.
67. SPO, CSORP 1889/5670.
68. *FJ,* 24 September, 1, 8, 15, 22, 29 October 1888.
69. SPO, CLPA Prosecutions.
70. *FJ,* 29 October 1888.
71. SPO, CSORP 1888/23576.
72. SPO, CLPA Prosecutions.
73. Ibid. For O'Brien's richly entertaining account, see *Evening memories,* pp.360-72.
74. *The Union,* 23 February 1889.
75. SPO, CLPA Prosecutions.
76. TCD Ms. 6560, John Dillon Papers, journal entry, 4 February 1889.

77. Ibid., 6836, O'Brien to Dillon, 5 February 1889.
78. *CC,* 15 February 1889.

Chapter 7: Deus Ex Machina

1. Salisbury Mss, correspondence with Sir Michael Hicks Beach, 1871-86, Hicks Beach to Salisbury, 15 December 1886.
2. Ibid., correspondence with Arthur Balfour, 1888-9, Balfour to Salisbury, 28 December 1888 and 2 November 1889. See also BL, Balfour Papers, Add. Ms.49826, Balfour to Ridgeway, 13 March 1888.
3. PDA Reports, 1881-7, NLI, Ir 3330941 P7.
4. The Land Corporation of Ireland Ltd., NLI P3000 and JP 346.
5. NLI Ms.22461 (vi, vii, viii), The O'Grady Papers.
6. The Anti-Plan of Campaign Association was founded in July 1887. According to *The Union,* 16 July 1887, 'it will combine owners, mortgagees, banks and insurance companies, all in fact who have any interest in Irish land'. It may have been administered as a branch of the Land Corporation.
7. See above pp.47-8.
8. NLI Ms. 22461 (vi, vii, x), The O'Grady Papers.
9. *ETC,* 16180-97.
10. *FJ,* 24 May 1889, Masserene to the editor, 23 May 1889.
11. *The Times,* 10 April 1889, G. Shaw Lefevre to the editor, 8 April 1889.
12. NLI Ms.13500 (4), T. P. Gill Papers. Henry McGee to Gill, 13, 14 April 1889.
13. *FJ,* 30 April 1889.
14. BL, Balfour Papers, Add. Ms.49809, Ridgeway to Balfour, 30 April 1889.
15. *The Times,* 16 March 1889.
16. Ibid., 8 April 1889.
17. *FJ,* 3 May 1889. See also PRO, CO 903/3, Irish Crime Records, 1892-94, Miscellaneous notes M, Series XV, October 1893, 'The Irish Derelict Land Trust Fund and the Brooke estate, County Wexford'.
18. *FJ,* 4 December 1889, Captain E. C. Hamilton to the editor.
19. SPO, CSORP 1889/14256.
20. PRO, CO 903/3.
21. *ETC,* 12889-97.
22. PRO CO 903/3.
23. *ETC,* p.14.
24. Cork Defence Union, *Boycotting in County Cork* (Cork, 1886).
25. *CE,* 28 September 1885.
26. *CC,* 13 December 1886.
27. Ibid., 16 December 1886.
28. *The Times,* 4 March 1887.
29. *FJ,* 7 March 1887.
30. *CC,* 16 April 1890.
31. *The Union,* 19 March 1887. See also Hansard, HC debs, 3S, Vol. 310, Col. 1162, 10 February 1887.
32. Salisbury Mss, correspondence with Arthur Balfour, 1887, Balfour to Salisbury, 27 November 1887.
33. Ibid., 1888-9, Balfour to Salisbury, 29 February 1888. For details of the Burke settlement, see Appendix 2, 'Plan of Campaign Estates'.
34. PRO, Cabinet Minutes, CAB (1889) 37/23, 'Political conditions of Ireland', Arthur James Balfour, 20 February 1889. See also CAB (1888) 37/22, No. 33, Balfour, 7 November 1888; *The Times,* 5 November 1887, Balfour at Birmingham, 4 November 1887 and Hansard, HC debs, 3S, Vol.322, Cols 739-40, 17 February 1888.

35. SPO, P and CR, 1887-1917, South Western and Western Divisions, December 1888.
36. Salisbury Mss, correspondence with Arthur Balfour, 1888-9, Balfour to Salisbury, 28 December 1888.
37. PRO, CAB (1889) 37/23.
38. SPO, CLPA Prosecutions.
39. The estates were those of De Freyne, Ellis, Fitzgerald, Gibbons, O'Hara, Pollock, Twiss, Townshend, Col Vandeleur and Worthington. See Appendix 2, 'Plan of Campaign Estates'.
40. BL, Balfour Papers, Memoranda on Clanricarde estate by A. J. Balfour, 7 November 1888 and 30 January 1889. See also Add. Ms.49827, Balfour to Ridgeway, 10 December 1888, 14 February 1889. See Ridgeway Papers, Balfour to Ridgeway, 15 December 1888, for a copy of Balfour's letter to Clanricarde, 12 December 1888.
41. See above p.32.
42. PRO, CO 903/1, Irish Crime Records, 1885-92, Intelligence Notes XII, 17 June 1889, 'A memorandum on the Ponsonby estate'.
43. *ETC,* 1156-91; 1627-31. See also Very Rev. D. Canon Keller PP, *The struggle for life on the Ponsonby estate: a chapter from the history of Irish landlordism* (London, 1887), pp. 12-15.
44. PRO, CO 903/1. See also The Liberal Union of Ireland, *The Plan of Campaign: the Ponsonby estate* (London 1890), p.4.
45. Salisbury Mss, correspondence with Arthur Balfour 1887, Balfour to Salisbury, 27 November 1887.
46. ETC, 1309-47.
47. SPO, CSORP 1889/7025.
48. Salisbury Mss, Correspondence with Arthur Balfour 1888-9.
49. BL, Balfour Papers, Add. Ms.49827, Balfour to Lord Northbrook, 21 January 1889, Viscount De Vesci, 25 January 1889, the Earl of Pembroke, 8 February 1889. It is unlikely that the former Under Secretary, Sir Redvers Buller, would have agreed with the inclusion of Viscount De Vesci in such a highly confidential and elaborate scheme. 'Lord De Vesci is a real good fellow and tries and does his best, but his brains are slow', he informed Sir Joseph West Ridgeway on 15 November 1887. See Ridgeway Mss.
50. BL, Balfour Papers, Add. Ms.49827.
51. *CC,* 15 February 1889. For a decidedly less complimentary account of the dinner, see *CE,* 16 February 1889.
52. *ETC,* 1309-47.
53. SPO, CSORP 1889/7025.
54. Irish Land Commission 1437, Final Schedule of Incumbrances, 406.
55. BL, Balfour Papers, Add. Ms.49814, Balfour to Peter O'Brien, 3 August 1889.
56. *FJ,* 20 July 1889.
57. *CC,* 27 January 1890. The report described him as a man 'of immense wealth, amounting to several millions sterling.' He owned no land in Ireland but was involved with several landlord organisations.
58. Hansard, HC debs, 3S, Vol.333, Col. 1504.
59. Ibid., Vol.337, Cols 1200-1201.
60. See, for example, *CE,* 23 March 1889, W. J. Lane to the editor of the *Evening Post,* 19 March 1889 and *ETC,* Appendix C, pp.446-9, Canon Keller to the editor of *The Times,* 1, 21 August 1889.
61. Hansard, HC debs, 3S, Vol.337, Cols 489-490, 21 June 1889. See also *The Times,* 17 August 1889, Smith-Barry to the editor, 15 August 1889.

62. *CE,* 28 May 1889 (from the *Hunts County News).*
63. SPO, CSORP 1889/7025.
64. *The Times,* 5 August 1889, Keller to the editor, 1 August 1889.
65. *ETC,* 1379-1401.
66. Ibid., 1402-20, 1436-7, 1445.
67. *FJ,* 11-16 March 1889, H. H. Townsend to the editor, 11, 14 March, W. J. Lane to the editor, 8 March, M. J. Horgan to the editor, 13, 15 March.
68. See *CE,* 19-24 June 1889 for details.
69. Ibid., 25 June 1889.
70. Hansard HC debs, 3S, Vol.337, Cols 473-482 and 491-2, 21 June 1889.
71. *FJ,* 24 June 1889.
72. Ibid., 25 June 1889.
73. Cork Archives Council, 'Cork union minute book, No. 88, March-September 1889'.
74. *FJ,* 28 June 1889, Croke to Very Rev. R. Cahill PP, VG, Tipperary, 26 June 1889.
75. Ibid.
76. PRO, CO 903/1, Irish Crime Records 1885-92, Intelligence Notes XIV, January 1890, 'Ponsonby and Smith-Barry estates (i) to 30 November 1889'.
77. *CE,* 29 June 1889.
78. Ibid., 1 July 1889, *UI,* 6 July 1889.
79. Ibid., 1, 2 July 1889.
80. BL, Balfour Papers, Add. Ms.49828, Balfour to Ridgeway, 23 July 1889.
81. *CE,* 2 July 1889.
82. *FJ,* 4 July 1889.
83. *CC,* 4 July 1889. See also *CC,* 8 July 1889, Robert Pratt of Carrigohane to the editor, 6 July 1889.
84. *DE,* 4 July 1889.
85. *CE,* 8 July 1889.
86. Ibid., 18 July 1889.
87. *FJ,* 26 July 1889.
88. Ibid., 5, 10 August 1889.
89. For details of the Tipperary struggle, see Marnane, 'Tipperary Town and the Smith-Barry estate' and the same author's, 'Fr David Humphreys and New Tipperary' in William Nolan (ed), *Tipperary: history and society. Interdisciplinary essays on the history of an Irish county* (Dublin, 1985), pp.365-378.
90. *Tipperary Nationalist,* 14 August 1889.
91. *UI,* 26 October 1889.
92. *CE,* 23 September 1889.
93. *Dublin Evening Mail,* 15 November 1889.

Chapter 8: The Tenants Defence Association

1. TCD Ms.6836, John Dillon Papers, O'Brien to Dillon, 14 July 1889.
2. Ibid., 6500, 'Irish parliamentary party minute book, 1886-1890', Vol.1. See also *FJ,* 16 July 1889. For the constitution of the Tenants Defence League, or the Tenants Defence Association as it was subsequently called, see NLI Ms. 21910, Heffernan Papers. See also *FJ,* 23 July 1889.
3. Ibid., 6836, O'Brien to Dillon, 14 August 1889.
4. NLI Ms. 13,500 (7), T. P. Gill Papers, J. J. Clancy to Dr [Kenny?], 2 September 1889.
5. *FJ,* 9 September 1889.
6. NLI Ms. 13,506 (1), Gill Papers, Sexton to Gill, 7 September 1889.
7. Ibid., 13,500 (7), W. J. Lane to Gill, 7 September 1889.
8. Ibid., 13,506 (11), Redmond to Gill, 4 October 1889.

9. *FJ*, 7 October 1889, Parnell to Sexton, 4 October 1889.
10. TCD Ms. 6836, Dillon Papers, O'Brien to Dillon, 21 October 1889.
11. NLI Ms. 10,946, John J. Dunne Papers, Matthew M. Kenny to Dunne, 16 November 1889.
12. *FJ*, 25 October 1889.
13. Ibid., 29 October 1889.
14. Ibid., 24 January 1890.
15. Ibid., 18 December 1890.
16. SPO, CBS, DICS, Midland Division, No. 4, 1887-1894, 5 December 1889.
17. *Archivium Hibernicum*, XXXII, 1974, Kirby Papers, McEvilly to Kirby, 29 December 1889 and *FJ*, 6, 13, 20, 27, 28, November 1889, 6, 11, 12, 13, 18 December 1889, 8, 10, 15, 16 January 1890.
18. SPO, CBS, DICS, Midland Division, No. 4, 1887-1894, 5 December 1889.
19. PRO, CO 903/1, Irish Crime Records, 1885-1892, Intelligence Notes VII, 'Memorandum on the rise and progress of the Tenants Defence Association, up to 31 January 1890'.
20. William O'Brien Papers, UCC, Box AC, Gill to O'Brien, 28 May 1890.
21. *FJ*, 26 May 1890.
22. Ibid., 30 May 1890.
23. *UI*, 2 August 1890.
24. William O'Brien Papers, UCC, Box AC, Gill to O'Brien, 28 May 1890.
25. TCD Mss 5945-8, 'The Smith-Barry estate case'.
26. *FJ*, 28 December 1889.
27. Ibid., 14 April 1890. See also Marnane, 'Tipperary Town and the Smith-Barry estate'.
28. BL, Balfour Papers, Add. Ms. 49811, Ridgeway to Balfour, 1 June 1890.
29. Croke Papers, originals in the possession of the Archbishop of Cashel, Archbiship's House, Thurles; microfilm copy in NLI, Smith-Barry to Croke, 8 May 1890. See also *CC*, 19 August 1891.
30. TCD Ms. 6731, John Dillon Papers, Ginnell to Dillon, 28, 29 January 1891, 10 August 1891.
31. William O'Brien Papers, UCC, Box AC.
32. *FJ*, 30 May 1890.
33. Hansard, HC debs, 3S, Vol.346, Cols 1498-9.
34. *FJ*, 14 July 1890.
35. SPO, CBS, DICS, Midland Division, No.4, 5 August 1890.
36. *FJ*, 25 August 1890.
37. NLI Ms. 8,574, T. C. Harrington Papers, Dillon to Harrington, 15 July 1890.
38. *FJ*, 25 August 1890.
39. Ibid., 26 August 1890.
40. *Archivium Hibernicum*, XXXII, 1974, Kirby Papers, Coffee to Kirby, 20 August 1890.
41. Lyons, *Parnell*, pp.420-3.
42. *FJ*, 17 September 1890, *Dublin Evening Mail*, 29 September 1890.
43. *Archivium Hibernicum*, Kirby Papers, Walsh to Kirby, 30 July 1890, Croke to Kirby, 3 August 1890.
44. NLI Microfilms, Laurence Gillooly Papers, Walsh to Gillooly, 17 August 1890.
45. *FJ*, 16 April-1 May 1890.
46. TCD Ms. 6754, John Dillon Papers, Kennedy to Dillon, 20 August 1890.
47. *FJ*, 15 September 1890.
48. PRO, CO 903/1, Irish Crime Records, 1885-1892, Intelligence Notes, XV.

49. *FJ,* 7 October 1890.
50. Ibid., 11 October 1890.
51. Lyons, *Dillon,* pp.111-113.

Chapter 9: Fin de Siècle

1. NLI Ms.8,574 (14), T. C. Harrington Papers, Dillon to Harrington, 18 December 1890.
2. TCD Ms.6732a, John Dillon Papers, Harrington to Dillon, 31 December 1890.
3. NLI Ms.13,500 (10), T. P. Gill Papers, Ginnell to Matthew Bodkin, 28 December 1890.
4. TCD Ms.6735, John Dillon Papers, J. F. X. O'Brien to Dillon, 27 December 1890.
5. See, for instance, *FJ,* 7, 19 January 1891 and SPO, P and CR, CBS, Western Division, report for February 1891.
6. Parnell stated at Listowel on 13 September 1891 that the Paris funds amounted to £46,000, *FJ,* 14 September 1891.
7. William O'Brien Papers, UCC, Box AD, receipt signed by Tim Harrington, 24 January 1891 and TCD Ms.6731, John Dillon Papers, Ginnell to Dillon, 29 January 1891.
8. PRO, CO 903/2, Irish Crime Records, 1887-1892, Miscellaneous Notes, Series XII.
9. SPO, P and CR, CBS, Western Division, report for December 1890 and DICS, Midland Division, report for January 1891.
10. Hansard, HC debs, 3S, Vol.349, Cols 1404-15, 30 January 1891.
11. *FJ,* 24 October 1887. See also Archbishop W. J. Walsh, 'Arbitration or the battering-ram', in the *Contemporary Review,* Vol.LV, June 1889, pp.797-814.
12. SPO, CLPA Prosecutions.
13. BL, Balfour Papers, Add.Ms.49827, Balfour to Ridgeway, 1, 4 April 1889.
14. Walsh, 'Arbitration'.
15. BL, Balfour Papers, Add.Ms.49827, Balfour to Ridgeway, 15 May 1889. The settlement of agrarian disputes by arbitration was not, as one authority has argued, part of government policy. See Curtis, *Coercion and conciliation,* p.242. As we have seen, the government, in the person of its Irish Chief Secretary, was completely opposed to the idea.
16. Hansard, HC debs, 3S, Vol.349, Cols 1484-96, 30 January 1891.
17. *The Times,* 31 January 1891.
18. TCD Ms.6754, John Dillon Papers, Fitzgibbon to Dillon, 23 January, 2 February 1891.
19. *FJ,* 7 January 1891.
20. TCD Ms.6754, John Dillon Papers, Fitzgibbon to Dillon, 5 February 1891.
21. SPO, P and CR, CBS, Western Division, reports for December 1890 and February 1891.
22. PRO, CO 903/2, Irish Crime Records, 1887-1892, Miscellaneous Notes, Series XIII, June 1891.
23. SPO, P and CR, CBS, and CBS, DICS, South-West Division, reports for April 1891.
24. SPO, CSORP 1893/3904.
25. *FJ,* 22, 27 April 1891, Redmond at a meeting of the central branch of the National League, Parnell at Clonmel, 26 April 1891.
26. Ibid., 12 August 1891, Tim Harrington at a central branch meeting.
27. Ibid., 26 January 1892, John Dillon to the editor, 23 January 1892.
28. *CC,* 1 June 1891.
29. Ibid., 3 October 1891, Smith-Barry to the editor, 29 September 1891.
30. *ETC,* 1446-64. See also *ETC,* Appendix C, Paper vii, pp.451-2.

31. TCD Ms.6736, John Dillon Papers, F. McGuire, Herbertstown to Fr Matthew Ryan, 29 January 1892; NLI Ms.22,461 (vi), The O'Grady Papers.
32. *ETC,* 13840–9.
33. *FJ,* 26 January 1892. John Dillon to the editor, 23 January 1892.
34. *ETC,* Appendix V, Table E.
35. HC 1907 [C.3570] LXXX.1155. *Irish Land Act, 1903 (Evicted Tenants). Special report of the estates commissioners of the proceedings up to 31 May 1907 in respect of persons applying as evicted tenants, or their representatives, for restoration to their former holdings, or for parcels of land in lieu thereof under the Irish Land Act, 1903.* The report pointed out that there were eight unsettled estates on 10 April 1905; but the Plan of Campaign had never been worked on the Cloncurry property.
36. Ibid.
37. Irish Land Commission, ETA 6777.
38. Finnegan, 'Land war in south-east Galway', p.133. For O'Brien's description of Clanricarde, see his introduction to Thomas Brett, *Life of the Most Reverend Dr Patrick Duggan, Bishop of Clonfert* (NP, 1921).

Conclusion

1. Michael Davitt, 'The evicted tenants problem', in the *Nineteenth Century,* Vol.XXXV, No.CCVI, April 1894, p.565.
2. *FJ,* 21 May 1889.
3. SPO, CSORP 1888/19903.
4. *FJ,* 10 January 1889.
5. HC 1876 (412), LXXX, 395-580. *Copy of a return of the names of proprietors and the area and valuation of all properties in the several counties of Ireland held in fee or perpetuity, or on long leases at chief rents.* Quoted in T. W. Moody, *Davitt and Irish revolution* (Oxford, 1981), p.560.
6. *ETC,* 14,955.
7. SPO, CBS, 1892, 4697/S, 'The Plan of Campaign, disposition and whereabouts of "the war-chest" ', May 1892.
8. *The Lansdowne Estate and the Matthew Commission. A statutory declaration drawn up by William Rochfort on 13 March 1893, refuting some of the evidence given before the Evicted Tenants Commission,* p.59.
9. NLI Ms.22461 (VI), The O'Grady Papers.
10. Mortgage ledgers of the Representative Church Body.
11. HC 1890, LX.74. *Land Court (Ireland) Estates. Return showing the number of properties under the control of the Land Court of Ireland, giving the dates when each property came under their control, the number of tenants and the present gross rentals of each estate where receivers have been appointed, and showing also when each estate was last put up for sale by the court.*
12. ILPU, *Some further "Notes and Queries" on the Irish question for the reflective voter,* (Dublin, 1887), p.4; Hansard, HC debs, 3S, Vol.310, Col.225, 28 January 1887; *The Times,* 9 December 1886.
13. *The Times,* 9, 20 December 1886.
14. Standish O'Grady, *Toryism and the tory democracy* (London, 1886), p.239.

Bibliography

MANUSCRIPT SOURCES

1 DUBLIN

State Paper Office

Chief Secretary's Office Registered Papers, 1886–1891.
Police and Crime Records, 1886–1917, Crime Branch Special. .
Irish Crime Records, 1879–1893. Returns of Agrarian Outrages Reported to the Constabulary.
Irish Crime Records, 1887–1891. Prosecutions under the Criminal Law and Procedure Act, 1887.
Irish National League Papers, 1879–1889 (10 cartons).
Special Commission, 1888 (7 cartons).

National Library of Ireland

Croke Papers (microfilm of originals in Cashel Diocesan Archives, Thurles): n 5710–13.
TS Catalogue by Dom Mark Tierney.
Timothy C. Harrington Papers: 'National League Letter Book, 1883–1888'.
Also Ms.8574.
George C. Mahon Papers: Ms.22231.
The O'Grady Papers: Ms.22461 (Xerox of originals in possession of The O'Grady, Kilballyowen, Bruff, County Limerick).
T. P. Gill Papers: Mss 13500, 13506.
John J. Dunne Papers: Ms.10946.
Laurence Gillooly Papers (microfilm of originals in Elphin Diocesan Archives, Sligo).

Library of Trinity College

John Dillon Papers.
Michael Davitt Papers.
Smith-Barry estate case: Mss 5945–8.

Representative Body of the Church of Ireland, Church House, Rathmines.

Mortgage ledgers.

Irish Land Commission

Final Schedule of Incumbrances.

2 CORK

University College

William O'Brien Papers.

Archives Council

Cork Union Minute Book, No.88, March–September 1889.

3 MAYNOOTH

St Patrick's College

Mortgage Records

4 BELFAST

Public Record Office, Northern Ireland

Colonel E. J. Saunderson Papers.

5 LONDON

Public Record Office

Buller Papers, WO 132/4A.
Cabinet Minutes, 1885-1889.
Irish Crime Records, 1885-1892, CO 903/1
1887-1892, CO 903/2
1892-1894, CO 903/3

British Library

Balfour Papers.

6 LEEDS

West Yorkshire Public Record Office

Clanricarde Papers

7 GLOUCESTER

Gloucestershire Record Office

Hicks Beach Papers

8 IN PRIVATE OWNERSHIP

William J. Duffy Papers – property of Desmond Duffy, Brewery View, Loughrea, County Galway.
Ridgeway Papers – property of Lord Tollemache, Helmingham Hall, Stowmarket, Suffolk, England.
Salisbury Papers – property of the Marquess of Salisbury, Hatfield House, Hatfield, Herts., England.

PRINTED RECORDS

PARLIAMENTARY PAPERS AND OFFICIAL PUBLICATIONS

1 LANDLORD TENANT RELATIONS

Return of owners of land of one acre and upwards in the several counties, counties of cities and counties of towns in Ireland [C.1492], HC 1876, LXXX. 35.

Report of the royal commission on the Land Law (Ireland) Act, 1881 and the Purchase of Land (Ireland) Act, 1885 [C.4969], HC 1887, XXVI, 1.

Report of the commissioners appointed to inquire into the estates of evicted tenants in Ireland [C.6935], HC 1893-1894, XXXI.

Irish Land Act, 1903 (Evicted Tenants). Special report of the estates commissioners of the proceedings up to 31 May 1907 in respect of persons applying as evicted tenants, or their representatives, for restoration to their former holdings, or for parcels of land in lieu thereof under the Irish Land Act, 1903 [C.3570], HC 1907, LXXX. 1155.

2 AGRARIAN CRIME

Return of the number of agrarian outrages committed in Ireland which were reported to the Inspector General of the Royal Irish Constabulary during the quarter ended 31 December 1886 [C.4955], HC 1887, LXVIII. 33; *for the quarter ended 31 March 1887* [C.5030], HC 1887 LXVIII. 37.

3 AGRICULTURAL STATISTICS

HC 1886 [C.4802], p.12, Table VIII; HC 1890-1891, XCI; HC 1892 [C.6777], CXXXVIII; HC 1893 [C.7187], p.17, Table VIII.

4 LEGAL PROCEEDINGS

Wilfrid Scawen Blunt v John Byrne in the High Court of Justice in Ireland [C.5401], HC 1888, LXXXIII. 11.

Criminal Law and Procedure (Ireland) Act 1887. Proceedings against members, HC 1889 [158] LXI. 515.

Special Commission Act, 1888: reprint of the shorthand notes of the speeches, proceedings and evidence taken before the commissioners appointed under the above-named act. 12 Vols, London, 1890.

Return of inquiries held under section one of the Criminal Law and Procedure (Ireland) Act, 1887. HC 1893-1894 [215] LXXIV. Part II.

5 PARLIAMENTARY RECORDS

Hansard's parliamentary debates, 1886-1891.

CONTEMPORARY NEWSPAPERS AND MAGAZINES

Newspapers

Cork
Cork Constitution
Cork Examiner

Dublin (Daily)
Daily Express
Dublin Evening Mail
Freeman's Journal
Irish Times
National Press

Dublin (Weekly)
The Irish Catholic
The Nation
The Union
United Ireland
Weekly Freeman

Magazines

Contemporary Review
Fortnightly Review
Murray's Magazine
New Review
Nineteenth Century
Westminster Review

PRINTED CONTEMPORARY WORKS

Pamphlets and Polemical Articles

Anon.
'Mad Tipperary', in *Murray's Magazine,* May 1890.
'The papal rescript', in *Westminster Review,* Vol.CXXIX, June 1888.

Clancy, J. J.
Mr Dillon and the Plan of Campaign. London, 1887.
'The position of the Irish tenant', in *Contemporary Review,* Vol.56, July 1889.

Davitt, Michael.
'The evicted tenants problem', in *Nineteenth Century,* Vol.XXXV, No.CCVI, April 1894.

Dillon, John.
'The coming winter in Ireland', in *Nineteenth Century,* Vol. XX, No. CXVII, November 1886.

Hamilton, Captain E.C.
The Woodford evictions. Dublin, 1886.

Irish Loyal and Patriotic Union.
Irish public opinion in support of the legislative union. Dublin, 1886
Social order in Ireland under the National League. Dublin, 1886.
The Plan of Campaign, illustrated: an account of the Massereene estate. Dublin, 1888.
The Plan of Campaign, illustrated: an account of the Ponsonby, Kingston, Lansdowne, O'Grady and Brooke estates. Dublin, 1888.
The Catholic Bishop of Limerick on the Glensharrold evictions. Dublin, 1890.

Keller, Canon Donal.
The struggle for life on the Ponsonby estate. London, 1887.
The Ponsonby estate and Mr Smith-Barry. A reply. London, 1889.

Liberal Union of Ireland.
The Plan of Campaign: the Ponsonby estate. February 1890.
The Plan of Campaign: the Smith-Barry estate. June 1890.
Irish landlords and tenants. March 1892.

Massingham, H. W.
The Gweedore hunt: a story of English justice in Ireland. London, 1889.

Norman, Henry.
Bodyke: a chapter in the history of Irish landlordism. London, 1887.

O'Brien, William.
'Are the Irish evicted tenants knaves?', in *New Review,* October 1892.

Pease, Alfred E., MP.
Glensharrold in 1888. London, 1888.

Pellew, George.
In castle and cabin, or talks in Ireland in 1887. New York and London, 1888.

Russell, T. W.
Disturbed Ireland: the Plan of Campaign estates. London, 1889.

Walsh, Archbishop W.J.
'Arbitration or the battering-ram?', in *Contemporary Review,* Vol.LV, June 1889.

Waterford, Marquess.
'Notes on the latest land programme', in *Contemporary Review,* Vol. CLIV, December 1889.

LATER WRITINGS BY CONTEMPORARIES

Blunt, Wilfrid Scawen.
The land war in Ireland, being a personal narrative of events. London, 1912.

Bodkin, M. McD.
Recollections of an Irish Judge: press, bar and parliament. London, 1914.

Davitt, Michael.
The fall of feudalism in Ireland. London and New York, 1904.

Healy, T. M.
Letters and leaders of my day. 2 Vols, London, 1928.

Kettle, Andrew J.
The material for victory: being the memoirs of Andrew J. Kettle. Ed. L. J. Kettle, Dublin, 1958.

O'Brien, R. Barry.
The life of Charles Stewart Parnell, 1846-1891. 2 Vols, London, 1889.

O'Brien, William.
Evening memories, being a continuation of Recollections by the same author. Dublin, 1920.

Pease, Sir Alfred E.
Elections and recollections. London, 1932.

Robinson, Henry.
Memories: wise and unwise. London, 1923.

Ross, Sir John.
The years of my pilgrimage. London, 1924.

Russell, T. W.
Ireland and the empire: a review, 1800-1900. London, 1901.

HISTORICAL WORKS

Baker, Michael.
Gladstone and radicalism: the reconstruction of liberal policy in Britain, 1885-1894. Hassocks, 1975.

Bew, Paul.
Land and the national question in Ireland, 1858-1882. Dublin, 1978.

Brett, Rev. Thomas.
Life of Patrick Duggan, Bishop of Clonfert. Dublin, 1921.

Clark, Samuel.
Social origins of the Irish land war. Princeton, 1979.

Clark, Samuel and Donnelly, James S. Jr (ed).
Irish peasants: violence and political unrest. Manchester, 1983.

Cooke, A. B. and Vincent, J. R. (ed).
Lord Carlingford's journal. Reflections of a cabinet minister, 1885. Oxford, 1971.

Corish, Patrick J. (ed).
'The Kirby Papers, Irish College, Rome. Guide to material of public and political interest, 1884-1894, with addenda, 1852-1878', in *Archivicum Hibernicum*, XXXII, 1974.

Crotty, R. D.
Irish agricultural production: its volume and structure. Cork, 1966.

Curtis, L. P. Jr.
'Incumbered wealth: landed indebtedness in post-Famine Ireland', in *The American Historical Review*, Vol.85, No.2, April 1980.

Coercion and conciliation in Ireland, 1880-1892: a study in conservative unionism. Princeton, 1963.

Donnelly, J. S. Jr.
The land and the people of nineteenth century Cork: the rural economy and the land question. London and Boston, 1975.

Drudy, P. J. (ed).
Ireland: land, politics and people. Irish Studies 2, Cambridge, 1982.

Dugdale, Blanche E.C.
Arthur James Balfour. 2 Vols, London, 1936.

Eversley, Lord.
Gladstone and Ireland: the Irish policy of parliament from 1850-1894. London, 1912.

Finnegan, A. B.
The land war in south-east Galway, 1879-1890. M.A., University College, Galway, 1973.

Fitzpatrick, David.
'Unrest in rural Ireland', in *Irish Economic and Social History,* Vol. XII, 1985.

Gallagher, J. P.
A life of Cardinal Patrick O'Donnell, 1857-1927. M.A., St Patrick's College, Maynooth, 1976.

Geary, Laurence M.
The land war on the Kingston estate, 1879-1888. M.A., University College, Cork, 1979.

Hammond, J. L.
Gladstone and the Irish nation. London, 1938.

Heyck, Thomas William.
The dimensions of British radicalism: the case of Ireland, 1874-1895. Urbana, Chicago, London, 1974.

Hicks Beach, Lady Victoria.
The life of Sir Michael Hicks Beach, First Earl St Aldwyn. 2 Vols, London, 1932.

Hyde, H. M.
The life of Sir Edward Carson. London, 1953.

Kirkpatrick, R. W.
Landed estates in mid Ulster and the Irish land war, 1879-1885. Ph.D., University of Dublin, 1976.

Larkin, Emmet.
The Roman Catholic Church and the Plan of Campaign, 1886-1888. Cork, 1978.

Livingstone, Peadar.
The Monaghan story: a documented history of the County Monaghan from the earliest times to 1976. Enniskillen, 1980.

Locker-Lampson, G. L. T.
A consideration of the state of Ireland in the nineteenth century. London, 1907.

Lyons, F. S. L.
The Irish parliamentary party, 1890-1910. London, 1951.

The fall of Parnell, 1890-1891. London, 1960.

'John Dillon and the Plan of Campaign, 1886-1890', in *Irish Historical Studies,* XIV, No.56 (September, 1965).
John Dillon: a biography. London, 1968.
Charles Stewart Parnell. London, 1977.

Lyons, F. S. L. and Hawkins, R. A. J. (ed).
Ireland under the union: varieties of tension. Essays in honour of T. W. Moody. Oxford, 1980.

MacDonagh, Michael.
The life of William O'Brien. London, 1928.

Marnane, Denis.
Tipperary town and the Smith-Barry estate, 1885-1895. M.A., University College, Dublin, 1973.

Marsh, Peter.
The discipline of popular government: Lord Salisbury's domestic statescraft, 1881-1902. Sussex and New Jersey, 1978.

Melville, C. H.
Life of General the Right Hon. Sir Redvers Buller. 2 Vols, London, 1923.

Miller, D. W.
Queen's rebels: Ulster loyalism in historical perspective. Dublin and New York, 1975.

Morley, John.
The life of William Ewart Gladstone. 3 Vols, London, 1903.

Murphy, Desmond.
Derry, Donegal and modern Ulster, 1870-1921. Derry, 1981.

Newton, Lord.
Lord Lansdowne: a biography. London, 1929.

Nolan, Peter.
J. F. X. O'Brien, MP (1828-1905). M.A., University College, Cork, 1975.

O'Brien, Conor Cruise.
Parnell and his party, 1880-1890. Oxford, 1957; corrected impression, 1964.

O'Brien, J. V.
William O'Brien and the course of Irish politics, 1881-1918. University of California Press, 1976.

O'Brien, Terence H.
'Lord Milner's Irish journal, 1886', in *History Today,* Vol.XIV, January 1964.

Ó Broin, Leon.
The prime informer: a suppressed scandal. London, 1971.

O'Neill, Brian.
The war for the land in Ireland. London, 1933.

Palmer, N. D.
The Irish Land League crisis. New Haven, 1940.

Pomfret, John E.
The struggle for land in Ireland, 1800-1923. Princeton, 1930.

Shaw, George Bernard.
The matter with Ireland, edited with an introduction by David H. Greene and Dan H. Lawrence. London, 1962.

Solow, Barbara L.
The land question and the Irish economy, 1870-1903. Cambridge (Mass.), 1971.

Tierney, Mark.
Croke of Cashel: the life of Archbishop Thomas William Croke, 1823-1902. Dublin, 1976.

Townshend, Charles.
Political violence in Ireland: government and resistance since 1848. Oxford, 1983.

Walsh, P. J.
William J. Walsh, Archbishop of Dublin. Dublin, 1928.

Woods, C. J.
The Catholic Church and Irish politics, 1879-1892. Ph.D., University of Nottingham, 1968.

Index